WORLD PHILOSOPHY SERIES

PHILOSOPHY OF SCIENCE

World Philosophy Series

Additional books in this series can be found on Nova's website under the Series tab.

Additional E-books in this series can be found on Nova's website under the E-book tab.

WORLD PHILOSOPHY SERIES

PHILOSOPHY OF SCIENCE

SAMUEL PINTUCK
AND
COLIN REYNOLDS
EDITORS

Nova Science Publishers, Inc.
New York

For permission to use material from this book please contact us:
Telephone 631-231-7269; Fax 631-231-8175
Web Site: http://www.novapublishers.com

NOTICE TO THE READER

LIBRARY OF CONGRESS CATALOGING-IN-PUBLICATION DATA

Philosophy of science / editors, Samuel Pintuck and Colin Reynolds.
p. cm.
Includes index.
ISBN 978-1-62100-276-5 (hardcover)
1. Science--Philosophy. I. Pintuck, Samuel. II. Reynolds, Colin.
Q175.3.P48 2011
501--dc23

2011033357

Published by Nova Science Publishers, Inc. † New York

CONTENTS

PREFACE

The philosophy of science is concerned with the assumptions, foundations, methods and implications of science. It is also concerned with the use and merit of science and sometimes overlaps metaphysics and epistemology by exploring whether scientific results are actually a study of truth. In this book, the authors present current research in the study of the philosophy of science including explaining global interactions; Poincare's enigmatic relationship with logic and the infinite; providing analogous insight regarding ontological issues and a conceptual parallel between Heisenberg's and Bohm's interpretation of quantum mechanics.

Chapter 1 - Several variants of Benardete's paradox lead naturally to the existence of global interactions and there are also other variants that cannot be understood in terms of global interaction because they are examples of actions with no interaction at all. The model of Benardete's paradox employed in this chapter enables us to see how both global interaction and action without interaction can emerge as different manifestations of the same explanatory schema, obeying to that extent a common "deep reason". Finally, in the light of these results, I shall consider Shackel's criticism (2005) of several current diagnoses concerning Benardete's paradox.

Chapter 2 - Traditionally, Henri Poincaré (1845 - 1912) has been recognized as a pioneer in the discovery of predicative set theory. Less attention, however, has been attributed to the importance of intuition to his interpretation of what it means to exist in the domain of pure mathematics. This imbalance is hard to justify in terms of Poincaré's own philosophical writings, as his contentions with his contemporaries – most notably, the logicists – are typified by appeals to the indispensability of intuition as a foundation for mathematics. The value of predicativism as a means of circumventing the set-theoretic paradoxes and indeed, of rendering mathematical definitions and proofs more transparent is well-recognized. Nevertheless, the associated systems are also noted for their restrictive nature in terms of proof-theoretic strength. It is important, therefore, to take a fresh look at Poincaré's philosophical writings to establish how the divergence between his foundational arguments and his reputation as a predicativist may have emerged. In this chapter, I shall offer new insight into the relevance of intuition to Poincaré's criteria for set existence and in particular, his requirement that infinite sets should be surveyable. On this basis, I shall argue that Poincaréan set theory ought to be more liberal than has traditionally been assumed. In exploring the above criteria, I shall also revisit Poincaré's unexplained affinity towards classical logic. By appealing to the influence of Aristotle's writings on Poincaré's interpretation of logic, I shall show that his ambiguous reference to the intuition of pure number as that of pure logical forms invites further exploration. Thus, I shall argue that Poincaré's treatment of the principles of Non-contradiction and Excluded Middle as necessarily true has its roots in logical intuition – a variety of intuition which he neglects at the cost of his own anti-logicist stance. These findings serve as an extension of my recent work on the nature of mathematical intuition. As such, they provide some essential groundwork for developing an alternative set theory with the potential to create a more successful synergy between Poincaré's philosophy and practice in pure mathematics.

Chapter 3 - In this chapter, the authors argue that both the problem of quantum indistinguishability in the philosophy of physics and the problem of identity in structures in the philosophy of mathematics are identical in a certain relevant sense; i.e., that they are two instances of the same general problem. This favors the idea that certain ontological questions are common to different domains, in spite of the specificity of each domain; hence, the solutions proposed in one of the domains may be extrapolated and adapted to the others.

Chapter 4 - It is undeniable that quantum mechanics has enjoyed nothing less than staggering success in almost every way that scientific theories are judged. Although it is a theory whose laws govern the unimaginably small, it has had a profound impact on the everyday lives of people.

Chapter 5 - Recent research in science education has shown considerable interest in the history, philosophy of science and epistemological issues. This, relatively new approach to teaching and research in science education requires a brief recapitulation of the debates and controversies in history and philosophy of science.

In: Philosophy of Science
Editors: S. Pintuck and C. Reynolds
ISBN: 978-1-62100-276-5

Chapter 1

GLOBAL ACTION AND ACTION WITHOUT INTERACTION

Jon Perez Laraudogoitia
Logic and Philosophy of Science Department Universidad del País Vasco C/ Paseo de la Univesiad, Vitoria-Gasteiz, Spain

ABSTRACT

Several variants of Benardete's paradox lead naturally to the existence of global interactions and there are also other variants that cannot be understood in terms of global interaction because they are examples of actions with no interaction at all. The model of Benardete's paradox employed in this chapter enables us to see how both global interaction and action without interaction can emerge as different manifestations of the same explanatory schema, obeying to that extent a common "deep reason". Finally, in the light of these results, I shall consider Shackel's criticism (2005) of several current diagnoses concerning Benardete's paradox.

1. INTRODUCTION

In "A Variant of Benardete's Paradox" (2003), I showed that several variants of Benardete's paradox lead naturally to the existence of global interactions, and in "Action without Interaction" (2005) I showed that there are also other variants that cannot be understood in terms of global interaction because they are examples of actions with no interaction at all. The latter result in particular required postulating two things:

PW(B): There is no positive lower bound on the possible length of time that a change in the position of a particle may take,

and

C: Particle world lines do not have beginning or end points.

PW(B) is *de facto* incompatible with a relativist universe such as, for example, the one in which we live. It would therefore be desirable to have to hand an argument for the possibility of action without interaction that did not depend on this postulate. This is one of the things I shall be doing here: in the present article only the use of the usual postulates of continuity ("Particle world lines are continuous") and connectedness ("If a particle exists in two given instants of time it also exists in every instant of time between them") (1) will be crucial, although the latter postulate is not required to prove the existence of action without interaction. But the model of Benardete's paradox employed to do so also enables us to see how both global interaction and action without interaction can emerge as different manifestations of the same explanatory schema, obeying to that extent a common "deep reason". Finally, in the light of these results, I shall consider Shackel's criticism (2005) of several current diagnoses concerning the Benardete paradox, in particular the one given in my, "A Variant of Benardete's Paradox"

2. Explaining Global Interaction

In the model of Benardete's paradox I use in this chapter, I shall assume, for simplicity's sake, unidimensional motion (as is usual) and I shall assume that particle p is moving to $\Delta<t<0$ in the region $x<0$ of axis X according to the law $x = t$. Another assumption is that only the gods (a numerable infinity of gods: God_1, God_2, God_3 …) may affect p in one way or another. Following the model of the original Benardete's paradox, we will suppose that, for each n, God_n controls point $x_n = 1/n$ of space exclusively: this control consists in God_n being able to stop or not, at will, particle p at $x_n = 1/n$. The idea is that

God_n acts on the particle at $x_n = 1/n$ stopping it there or does not act on it at all at $x_n = 1/n$. This way, an individual God may not perform any action other than stopping the particle.[(2)]

Let us suppose that the action plan for God_n ($\forall n$) is as follows:

(1) God_n stops p at $x_n = 1/n$ if and only if no God_{n+m} ($m \geq 1$) stops p at $x_{n+m} = 1/(n+m)$ ($n = 1, 2, 3 \ldots$)

Clearly (1) leads to contradiction. Either no God stops p or at least one does so. If none stops it, then under (1), God_1 stops p at $x_1 = 1$, clearly a contradiction. If at least one God stops it, for example God_G, then under (1) no God_{G+m} ($m \geq 1$) stops it and, in consequence, no God_{G+m+1} ($m \geq 1$) stops it. But this last leads under (1) to God_{G+1} stopping the particle, another contradiction. My first contribution in this chapter (central to what follows) will be to explain why there is a contradiction here. There is a contradiction, as we have just seen; nevertheless I maintain that we may only say that we really understand the situation if we can explain why this is so, why (1) is defective. The key lies in seeing that in (1) only the more or less arbitrary condition that God_n imposes for acting (stopping p) is reflected, and that there is in fact another condition, one that does not depend at all on the will of God_n but (assuming, as we are, that God_n has the power to stop p at x_n) on what it actually means for p to stop at $x_n = 1/n$, and which is therefore the condition of the possibility of such a stop. To see this clearly, let us consider the sentence S: the individual I* stops particle p at the point x. S is equivalent to sentence A: the individual I* actualizes the possibility of particle p stopping at point x. Therefore a necessary and sufficient condition for S may be formulated as a conjunction of a necessary and sufficient condition for the possibility of particle p stopping at point x (assuming that I* has the power to stop p at x) with a condition necessary and sufficient for individual I* to actualize such a (now guaranteed) possibility. A necessary and sufficient condition for it to be possible for particle p to stop at point x is, fairly obviously, that the particle is at x in some moment t, such that there are instants of time previous, but arbitrarily close, to t in which the particle is at points different from x. This last is eminently reasonable because if, at least in the immediate past of t, the particle has never been at points different from x then it must have stopped before t (or perhaps never, for the trivial reason that it was never at points different from x). In more precise terms, it is possible for particle p to stop at point x if and only if it is possible for particle p to stop at point x in some instant t if and only if there is at least one instant t such that p is at x in t and $\forall \varepsilon > 0$ there is at least one instant belonging to the interval $(t - \varepsilon, t)$ in which p is at a point different from x. Also, a necessary and sufficient condition for the individual I* to actualize the (now guaranteed) possibility of particle p stopping at point x may take a wide variety of different (and not necessarily mutually equivalent) forms owing to the fact that it will contain in general the more or less arbitrary criterion that I* decides to follow. In the case of the model of Benardete's paradox introduced above, this necessary and sufficient condition for God_n is of course given by the term on the right in equivalence (1). The defect in (1) is now clear: the necessary and sufficient condition for the possibility of p stopping at the different $x_n = 1/n$ points has not been included there. The action plan of God_n ($\forall n$) must therefore be written as follows:

(2) God_n stops p at $x_n = 1/n$ if and only if : no God_{n+m} ($m \geq 1$) stops p at $x_{n+m} = 1/(n+m)$ and there is at least one instant t_n such that: p is at x_n at t_n and $\forall \varepsilon > 0$ there is at least one instant in the interval $(t_n - \varepsilon, t_n)$ in which p is at a point different from x_n

How does this affect our diagnosis of the situation? Given (2), the fact that (1) leads to contradiction indicates that it cannot be true that, for all n:

(3) there exists at least one instant t_n such that: p is at x_n at t_n and $\forall \varepsilon > 0$ there is at least one instant in the interval $(t_n - \varepsilon, t_n)$ in which p is at a point different from $x_n = 1/n$

It is in fact easy to see that (3) is false for any n. To prove this, it is enough to use the postulates of continuity and connectedness (these being intuitively plausible and, following what was said at the beginning, all that is required, besides the powers we have supposed for the gods, to demonstrate both the possibility of global interaction and action without interaction). In fact, if (3) were true for a certain n = N then p would be at $x_N = 1/N$ at a certain instant t_N. This implies that none of the God_{N+m} ($m \geq 1$) gods would have stopped the particle, otherwise this would have stayed at some $x_{N+m} =$

$1/(N+m)$ and would never have arrived at point $x_N = 1/N$. However, to get to $x_N = 1/N$, particle p has had to pass through the x_{N+m} points at certain t_{N+m} instants (a requirement of the postulate of connectedness as, to begin with, p was in the region $x<0$ and we know that its space-time trajectory must be continuous). But this means that for any $m \geq 1$ there is at least one t_{N+m} instant such that: p is at x_{N+m} at t_{N+m} and $\forall\ \varepsilon > 0$ there is at least one instant in the interval $(t_{N+m} - \varepsilon, t_{N+m})$ in which p is at a point different from x_{N+m}, Therefore, for any $m \geq 1$ we deduce from (2) that:

(2.2) God_{N+m} stops p at $x_{N+m} = 1/(N+m)$ if and only if no God_{N+m+q} $(q \geq 1)$ stops p at $x_{N+m+q} = 1/(N+m+q)$

Since no God_{N+m} stops p it follows that no God_{N+m+q} $(q \geq 1)$ stops it and, in line with (2.2), that all the God_{N+m} gods stop the particle. This contradiction shows that the assumption we started with, i.e. that (3) is true for a certain $n = N$, is false. So (3) is false for all n. Therefore, whatever n is:

(4) for all instants t_n, if p is at $x_n = 1/n$ at t_n then $\exists\ \varepsilon > 0$ such that for all instants belonging to the interval $(t_n - \varepsilon, t_n)$ p is not at a point different from $x_n = 1/n$

It is clear that if, after $t = \Delta<0$, p is ever at some point $x_n = 1/n$ $(n = 1, 2, 3, ,,,)$, it will be so for instants of time $t \geq 0$ because, under the hypothesis, it is in the region $x< 0$ for $\Delta<t<0$. And if $t^* \geq 0$ is an instant in which p is at some point $x_n = 1/n$ then, in any instant prior to t^* and after Δ, p must be at some point because, by virtue of the postulate of connectedness, if it exists for $\Delta<t<0$ and for a certain $t^* \geq 0$, it must necessarily also exist in the non-negative intermediate instants (if there are any). That said, it is clear from (4) that if p is actually at some non-negative time (i.e. in some $t_n \geq 0$) at some point $x_n = 1/n$ $(n= 1, 2, 3, ,,,)$ then it will remain there at rest for ever because not being at a point different from $x_n = 1/n$ during a certain non-null interval $(t_n - \varepsilon, t_n)$, and since we know it will be at some point in each instant of that interval, it follows that it will be at $x_n = 1/n$ during the interval $(t_n - \varepsilon, t_n)$, i.e. it will be at rest in $x_n = 1/n$ during the interval $(t_n - \varepsilon, t_n)$. But then it will remain at rest at $x_n = 1/n$ for ever because it will have been stopped there by God_n (the only one controlling this particular point) and we may suppose that if any God stops p at some point it keeps it there for ever. So, since if p is ever at $t_n \geq 0$ at some point $x_n = 1/n$ (n= 1, 2, 3, …) it will remain there for ever and that for $\Delta < t < 0$ p is in the region $x<0$, the postulate of connectedness leads to the conclusion that p can never be in the region $x>0$. Consequently p will be stopped at $x=0$ or, alternatively, will rebound backwards at it. This way we recover the idea of global interaction introduced in "A Variant of Benardete's Paradox" (2003): no God interacts individually with p, but the set of Gods does so globally by preventing p from reaching the region $x>0$. The gain here over what was done then is that there the idea of global interaction was a hypothesis that required justification to make it plausible. Here, to the contrary, it is the consequence of a much more fundamental requirement, that of also making explicit the conditions for the possibility for any act of detention performable by the gods, and not merely the conditions of actualization of that possibility as if the possibility itself were already guaranteed. Although the latter (i.e. the mere conditions of actualization) is sufficient in the immense majority of contexts that we can usually imagine (and certainly in all ordinary contexts), Benardete's paradox reminds us that it is not always so.

In fact, in Benardete's original paradox, the action plan for God_n (n = 1, 2, 3, …) is not given by (1) but by the following: God_n stops p at point $x_n = 1/n$ should p actually arrive there. However, even in this case, the above discussion remains fully pertinent and valid because (1) continues to provide the necessary and sufficient condition for God_n to actualize the possibility of p stopping at $x_n = 1/n$ ASSUMING THAT THIS POSSIBILITY IS ALREADY GUARANTEED. Therefore (1) is defective for the same reason as before: it takes no account of the necessary and sufficient condition for the possibility of p stopping at $x_n = 1/n$. When this condition is added, we arrive at (2) and now from here the existence of a global interaction follows (under the postulates of continuity and connectedness) as a logical consequence.(3)

Before concluding this section, I would like to comment briefly on the indeterminism involved in the evolution of particle p when I say that p will be stopped at $x = 0$ or, alternatively, will rebound backwards at that point. One might think this conclusion makes my model less interesting, for the following reason: as any given force must have a unique effect on a particle, classical physics would seem to be bound up with determinism, so that cases of indeterministic evolution (and my model presents one of these cases) would not be consistent with it. But this conclusion is erroneous. Although a given force should have a single effect when acting on a particle, this does not turn classical physics into a deterministic theory because there are situations where it is not determined which actual force acts on a particle. Consider for example the case of three identical rigid spheres in line: two of them, B and C, at rest and in contact while the third, A, approaches them from the left at unit velocity. Classical physics does not permit us to determine the final state of this process. A may be stopped while C moves to the right. But it is also possible for both B and C to move to the right while A rebounds to the left. Given the force that acts on A (which in this case is specified not by a function but by a Dirac δ distribution) we will know exactly how it will move, the problem is that classical physics does not tell us what force acts on A: it leaves open a range of possibilities, albeit fairly limited. So the indeterminism linked to the global interaction of the gods with particle p does not *per se* make my model inconsistent with classical physics. Indeed, if we use the model of God_n as a particle q_n identical to p and placed at rest at $x_n = 1/n$ before $t = 0$ (the model introduced in note (3)) then the analysis of the global collision in Pérez Laraudogoitia 2005b, pp. 327-330, shows that classical mechanics does not determine which of the two destinies p follows: whether it is stopped or rebounds. Further, both here and in the case of the spheres A, B, C either of the two destinies of p (or A) is compatible with the elastic nature of the collision (multiple in A's case and global for p): the elasticity of the collision neither excludes p (or A) rebounding nor excludes the possibility of it stopping and thus it is compatible with indeterminism (in both cases we might suppose, without being inconsistent, that all the collisions involved are elastic).

3. Unifying Global Interaction and Action without Interaction

Let us see now how the same explanatory schema that has enabled us to demonstrate the possibility of global interaction also enables us to demonstrate the possibility of action without interaction. The modifications needed to the previous proof tend to the trivial and it is particularly important that the only necessary postulate is the one concerning the continuity of a particle's worldline. The interest and the relevance of this argument, which does not require the use of the postulate PW(B), will be immediately appreciable. All we have to do is take the limit $n \to \infty$ in the model of Benardete's paradox considered in the previous section. This of course only affects the terms of that model that have metric value, so that fractions $1/n$ and $1/(n + m)$ are now substituted by 0 (as $1/n \to 0$ and $1/(n + m) \to 0$, given that $n \to \infty$). Contrariwise, terms such as God_n, God_{n+m}, x_n, x_{n+m} and t_n are simply denotative, being replaceable, in principle, by arbitrary proper names, which means they would not be affected. Solely for simplicity's sake, I shall write $x = 0$ instead of $x_n = 0$ or $x_{n+m} = 0$. In consequence, in the new model of Benardete's paradox there are only two basic differences with regard to the previous model. The first is that, for each n, God_n controls the point x=0 exclusively. The second, that the action plan for God_n ($\forall n$) is as follows:

(1*)God_n stops p at x=0 if and only if no God_{n+m} ($m \geq 1$) stops p at x=0 (n= 1, 2, 3, …)

As occurred with (1), (1*) leads likewise to contradiction, and following the same path that took us from (1) to (2) the action plan of God_n ($\forall n$) must be written as:

(2*) God_n stops p at x=0 if and only if : no God_{n+m} ($m \geq 1$) stops p at x=0 and there is at least one instant t_n such that p is at x=0 in t_n and $\forall \varepsilon > 0$ there is at least one instant in the interval $(t_n - \varepsilon, t_n)$ in which p is at a point different from x=0

Once again, the contradiction found from (1*) indicates that it cannot be true that, for all n:

(3*) there is at least an instant t_n such that: p is at x=0 in t_n and $\forall \varepsilon > 0$ there is at least one instant in the interval $(t_n - \varepsilon, t_n)$ in which p is at a point different from x=0

and now it is also easy (it is in fact trivial) to prove that (3*) must be false for any n. Indeed, let N be a value of n for which (3*) is false, i.e. that it holds that:

it is false that there is at least one instant t_N such that: p is at x=0 in t_N and $\forall \varepsilon > 0$ there is at least one instant in the interval $(t_N - \varepsilon, t_N)$ in which p is at a point different from x=0

But in this sentence t_N is a bound variable and the symbol N only appears in it as a subscript of t, which means that N could be replaced by any other term n denoting a number without altering its truth value or even its meaning. Therefore, whatever n is:

(4*) for any instant t_n, if p is at x=0 in t_n then $\exists\, \varepsilon > 0$ such that for all instants belonging to the interval $(t_n - \varepsilon, t_n)$ p is not at a point different from x=0

From (4*), through elimination of the universal:

if p is at x=0 in t=0 then $\exists\, \varepsilon > 0$ such that for all instant belonging to the interval $(- \varepsilon, 0)$ p is not at a point different from x=0.

Now let us suppose that p is at x=0 in t=0. Then in some non-null interval of time $(- \varepsilon, 0)$ p will not be at a point different from x=0. But this contradicts the initial condition that for $\Delta<t<0$ p is in the region x<0. So p cannot be at x=0 in t=0. Since, according to our initial condition, p moves according to the law x=t for $\Delta<t<0$, the postulate of continuity implies that p cannot be in t=0 at some point $x \neq 0$ either. It follows that in t=0 p cannot be at any point in space, and, as for $\Delta<t<0$ it was moving freely (in the region x<0) this means that p disappears suddenly at t=0 (this is what it means to say it cannot be anywhere in t=0) without having been disturbed by any interaction whatsoever.(4) We thus recover the idea of action without interaction introduced in "Action without Interaction" (2005): no god interacts individually with p or acts individually on p; neither does the set of gods interact globally with p, although it does act globally on it, without any interaction, by making it disappear in t=0. Part of the gain here over what was done then is the economy of postulates, continuity alone being required (without connectedness and, above all, without PW(B)). But now the action without interaction also obeys the same "deep reason" as the global interaction given above: the need to make explicit also the conditions of possibility of any act of detention performable by the gods and not merely the conditions for the actualization of that possibility as if the possibility itself were already guaranteed. Note that, in all this, the statement that in t = 0 p cannot be at any point in space does not express a mere possibility that I decide to choose from a range of alternatives incompatible with it, although possible as well. It is rather a strict logical consequence of (2*) together with the pertinent initial conditions (namely, that p moves according to the law $x = t$ for $\Delta<t<0$) and the postulate of continuity. Therefore p's disappearance at $t = 0$ (and, consequently, the fact that p neither passes point $x = 0$ nor is stopped there) is not an option we have arrived at following our intuitions but rather the conclusion of a deductive argument.

In my second model of Benardete's paradox, one objection may be raised to the effect that, since all the gods control the same point $x=0$, God_n cannot reliably condition whether he acts or not on p at a particular instant to what other $Gods_{n'}$ ($n' \neq n$) do (or do not do) in the same instant, unless he has some exceptional premonitory faculties. This is not a crucial difficulty. To begin with, we may suppose that the gods are different from each other as regards the magnitude of their power to stop p or not at $x = 0$ so that in the list God_1, God_2, God_3, … they are placed in rising order of power. This implies in particular that if two gods and only two gods want to do exactly the same thing at exactly the same time (namely, stopping p at $x = 0$ in $t = 0$), only the more powerful of the two will do it. Let us now suppose that the action plan for God_n ($n=1, 2, 3, \ldots$) is not given by (1*) but by the following: God_n proposes to stop p at point $x=0$ should p ever arrive there. Obviously, no god needs any premonitory faculty. However, the above discussion remains fully pertinent and valid because (1*) continues to provide the necessary and sufficient condition for God_n to actualize the possibility of p stopping at $x=0$ ASSUMING THAT THIS POSSIBILITY IS ALREADY GUARANTEED. Therefore (1*) is defective for the same reason as before: it takes no account of the necessary and sufficient condition for the possibility of p stopping at $x=0$. When this condition is added, we arrive at (2*) and now from here the existence of action without interaction follows (under the postulate of continuity) as a logical consequence.[(5)]

If in the conditions of the model of Benardete's paradox in this section 3 of the chapter, we suppose (we postulate) that particles are eternal (they are neither created nor destroyed), then p will reach $x = 0$ in $t = 0$ and it will therefore be possible to stop it at $x = 0$. But then this leads us to (1*), i.e. a contradiction. This contradiction indicates that the state of things posited is impossible. But note that in no case do we have here a logical contradiction, i.e. in no case are the gods trying to do something logically impossible (ultimately what each wants is to stop p at point $x = 0$ should p ever arrive there). The situation is more correctly described by saying that what the gods are trying to do is impossible because it is incompatible with the postulate that the particles are eternal, but it is not impossible because it is incompatible with logic (in other words, it is not logically impossible). It also is important to note that although the eternity of the particles is compatible with the principle of conservation of mass in classical particle dynamics, it is not actually required by that principle.[6] Finally, we can block the result of Benardete's paradox in this section 3 (even rejecting the postulate that particles are eternal) by postulating that p still exists at t=0 (it being possible to stop it at x=0 in t=0). But this would imply the circularity of rejecting the conclusion of the sudden disappearance of p at t=0 for the path of postulating that p still exist at t=0. We may also postulate that the possible sudden disappearance of p at t=0 has nothing to do with the gods trying or not to stop it (i.e. that the two circumstances are not causally related). But this would imply the circularity of rejecting the conclusion of the existence of a causal relation for the path of postulating that there is none. Although both paths are possible (although the second would require a slight reformulation of the principle of conservation of mass, see note (6)) , they veil an interesting possibility (remember that the model of Benardete's paradox in this section 3 is only the limit $n \rightarrow \infty$ of the model of Benardete's paradox analyzed in section 2) compatible with classical physics. Indeed, just as in note (3) I constructed a model in particle dynamics of Benardete's paradox introduced in section 2 of this chapter, one may also provide (although with some additional work) a purely mechanical model of Benardete's paradox introduced in this section 3 that is compatible with classical mechanics, as we shall now see:

3.1. A Model of Gods. Preliminaries

To begin, let us say that two rigid, extensive bodies are in dynamic contact at time t if and only if they are in contact at t and, should their velocities in t^- remain constant thereafter, they would overlap (interpenetrate) each other after t. In addition, any set of rigid, extensive bodies collides at time t if and only if each one is at t in dynamic contact with at least one of the others and its momentum in t^- is different from its momentum at t^+. One of the reasons for classical dynamics being a (partially) indeterminist theory is that, when A is at t in dynamic contact with two or more bodies (to simplify, I shall suppose throughout what follows that the other bodies are not in contact with each other) the theory does not enable us to determine the momentum exchanged with each one and, in consequence, it does not enable us either to determine which ones it collides with (it may do so with all of them or only with some and not with others). I shall construct here an extension, T, of classical dynamics (and, therefore, compatible with it) in which this class of indeterminism disappears when, moreover, all the (rigid and extensive) bodies involved have identical mass and the movement takes place in a sole spatial dimension (the specific postulates of T shall then be applicable in these particular circumstances). T shall be a theory of hidden variables. To that end, a one-to-one correspondence is established between all the (rigid and extensive, as we have supposed from the beginning) material bodies of identical mass and the set of real numbers, associating with each body a single number (hidden variable) that we will call its index. The way in which the introduction of indices eliminates dynamic indeterminism of the type mentioned is as follows. Our first postulate will guarantee that the only collisions ever will be binary:

I) If A collides with B in t then A does not collide in t with any C of an index greater than that of B

From I) it follows that if A collides with B at t then it will not collide either at t with any C with an index lower than that of B. But in the case where A at t is in dynamic contact with two or more bodies I) still leaves indeterminate which of them it will collide with. The second postulate makes explicit that the index of a material body is a measure of the 'power' it has to take part in a binary collision (the only type permitted by I) in the frame of T), so that a larger index means greater power. Supposing A is in dynamic contact with B at t:

II) If A does not collide with B at t then A collides at t with at least a C of an index greater than that of B

It might be thought that the idea of 'a larger index means greater power' is reflected much more intuitively in a postulate like:

I') A collides at t with the material object with the largest index with which it is in dynamic contact at t

But this would not function where A is in dynamic contact at t with a numerable infinity of other material objects, as in such a case there is none having the greatest index.

3.2. A Model of Gods. Action without Interaction

Let us suppose now that A is a disk of unit radius (and mass) moving at unit velocity in parallel to its axis of symmetry of revolution (which we shall take as the z axis) in direction towards the plane $z = 0$ at which is at rest a numerable infinity B_1, B_2, B_3, ..., B_n, ... of rings of unit mass and radii $r_n = n/(n+1)$ with a common centre in the origin of coordinates of the plane $z = 0$. How will this system evolve in the future? From I) and II) it follows that, supposing the index of B_n to be equal to its radius and A is in dynamic contact with B_n at t :

III) A collides with B_n at t if and only if A does not collide at t with any B_{n+m} $(m \geq 1)$ $(n = 1, 2, 3, \ldots)$

In fact, III) is strictly equivalent to the conjunction of I) and II). But the important thing is that it has the same structure as (1*), where the gods of increasing power to retain the particle p (which is now the object A) are the bodies B_n of increasing power to collide with A. This way, we have found in our theory of hidden variables T (which is an extension of classical dynamics) a purely dynamic model for the gods God_n: the B_n rings. Like (1*), III) leads to a contradiction. But this does not prove T to be inconsistent. As with (1*), III) provides the necessary and sufficient conditions for B_n to actualize the possibility of colliding with A ASSUMING THAT THIS POSSIBILITY IS ALREADY GUARANTEED. We must then, as was to be expected, add to III) the necessary and sufficient conditions for the possibility of B_n colliding with A at t, i.e., that A is in dynamic contact with B_n at t. So we have not III), but

IV) A collides with B_n at t if and only if : A does not collide at t with any B_{n+m} $(m \geq 1)$ and A is in dynamic contact with B_n at t

At $t = 0$, let A still be in the plane $z = 1$, moving at unit velocity towards the plane $z = 0$ for $0 \leq t < 1$. Particularizing IV) at time $t = 1$:

A collides with B_n at $t = 1$ if and only if A does not collide at $t = 1$ with any B_{n+m} $(m \geq 1)$ and A is in dynamic contact with B_n at $t = 1$

As, for at least one n, A is not in dynamic contact with B_n at t = 1, but for $0 \leq t < 1$ A was moving according to the law z = 1 – t, it follows that A cannot be in contact with B_n at t = 1. A therefore cannot be at t = 1 in z = 0. As occurs with the gods, we conclude (applying the postulate of continuity to each of the 'particles' that compose A) that at t = 1 A cannot be in any part of space, i.e. A disappears suddenly at t = 1 without having been disturbed by any interaction whatsoever. The B_n (our models of the gods God_n) have acted without interaction at t = 1.

It might be thought that T, by leading to the disappearance of the disk A, is incompatible with classical mechanics because, supposedly, classical mechanics will lead in this case to a final situation where the exchange of momentum between A and the B_n could be any compatible with the restrictions imposed by the conservation of momentum and energy, so that the theory of hidden variables would only have the liberty to choose which of these exchanges of momentum and energy actually takes place. However, this is not so. The system formed by A and the B_i has infinite mass, and, as I argue in note (13), in such a case the principles of the conservation of energy and momentum are not applicable, so that in fact these principles say nothing whatsoever about its final dynamic state. The sudden disappearance of A at t = 1 also takes place in a way compatible with C** (see note (6)): the world lines of the "particles" that compose A cannot be extended with continuity for $t \geq 1$ in any way compatible with the restrictions on A imposed by the theory T.

4. Stopping and Baking

The unifying schema presented in sections 2 and 3 of this chapter is suitably exemplified in the step from (1) to (2) (section 2, 'Explaining global interaction'). In (1) the necessary and sufficient conditions for God_n to actualize the possibility that p is stopped at $x_n = 1/n$ are specified and in (2) the necessary and sufficient conditions for the possibility for p to stop at $x_n = 1/n$ are added. If we call the latter H then what we did was pass from

(1) God_n stops p at $x_n = 1/n$ if and only if no God_{n+m} $(m \geq 1)$ stops p at $x_{n+m} = 1/(n+m)$

to

(2) God_n stops p at $x_n = 1/n$ if and only if (no God_{n+m} $(m \geq 1)$ stops p at $x_{n+m} = 1/(n+m)$ and H)

where, as noted, H expresses the necessary and sufficient conditions for the possibility of p stopping at $x_n = 1/n$, i.e. H is analytically equivalent to the sentence "it is possible for p to stop at $x_n = 1/n$" (it would be more correct to write H_n here instead of H given the evident dependence of the subscript n, which determines the stopping point. I shall not do so, however, because this precision is not necessary to understand the discussion that follows).

This unifying schema is generally applicable. Let, for example,

(A) God_n bakes the cake P if and only if no God_{n+m} $(m \geq 1)$ bakes the cake P

(A) leads to the same type of contradiction as (1). Our unifying schema would entail replacing (A) with

(B) God_n bakes the cake P if and only if (no God_{n+m} $(m \geq 1)$ bakes the cake P and H*)

where H* expresses the necessary and sufficient conditions for the possibility that cake P to be baked, i.e. it is analytically equivalent to the sentence "it is possible for cake P to be baked." But note that to apply the above correctly, H* in (B) must be a sentence analytically equivalent to the sentence "it is possible for cake P to be baked." If this crucial fact of my unifying schema is not taken into account, one will easily arrive at absurdities. For example, to "prove" that Shiva will create a round square, it would be enough to assume (B), replacing H* with the sentence "Shiva will not create a round square."

What significant difference is there between sentences (2) and (B) of this section? We have sufficiently clear physical intuitions, supported by classical physics, about particles being stopped (see for example note (3)) which have led us to accept not H (which follows from (2)) and, from there, to admit global interactions. But we have nothing even remotely similar as regards cake baking that would justify us accepting not H* (which follows from (B)), meaning that in this case we tend to say that the gods are trying to do something logically impossible. This is certainly not strictly correct: what the gods are trying to do is incompatible with the truth of H* and therefore impossible to the extent that H* is true (analogously to what we said in the previous section, here there is no liberty of choice or decision: what the gods are trying to do is expressed by (B), and as the conjunction of (B) and H* leads to a contradiction it follows that (B) is therefore impossible to the extent that H* is true) . But we tend to overlook this point because we unconsciously and implicitly assume (postulate) that H* is true, which means that the previous incompatibility seems to us to be a purely logical impossibility. Clearly, it all depends on the degree of plausibility of the postulates assumed in each case because of course the conjunction of (2) and H also leads to contradiction, from which it follows that (2) is impossible to the extent that H is true.

One last word on stopping and baking that will help to clarify still further the relation between the two. (2) formulates the necessary and sufficient conditions for God_n to stop p at $x_n = 1/n$ and (B) formulates the necessary and sufficient conditions for God_n to bake the cake P: both define the gods' plan of action in either case. In contrast, H formulates the necessary and sufficient conditions for the possibility that p is stopped at $x_n = 1/n$ and H* formulates the necessary and sufficient conditions for the possibility that P is baked: both dispense completely with any plan of action for the gods and are independent of it. Finally, (2) logically implies not H and (B) logically implies not H* because, as I said above, both the conjunction (2)&H and the conjunction (B)&H* lead to contradiction (in each case the path to contradiction clearly cannot dispense with the gods' plan of action, although H and H* do actually dispense with it). But in the case of particle p at least, not H, far from being incompatible with classical physics, actually leads as we have seen (with the aid, among other things, of (2), from which it was deduced) to the appearance of a global interaction. So the gods can follow action policy (2) and policy (2) is possible for them. (7)

At this point it would be useful to remember, as I made clear at the beginning of the chapter, that I suppose that $(\forall n)$ God_n only controls point x_n, and is capable of stopping particle p at $x_n = 1/n$ if p gets there. Should no God_n really have this capability (imagine specifically that no god is capable of altering the state of movement of p at x_n, whatever it might be) the situation changes radically. Until now H provided the necessary and sufficient condition for the possibility of p being stopped at $x_n = 1/n$, and to justify it we explicitly assumed that God_n was capable of stopping (had the capability to stop) particle p at $x_n = 1/n$ (see the discussion between (1) and (2) in section 2 of this chapter). If we do not make this assumption, then H must be more demanding and to provide the necessary and sufficient condition for the possibility of God_n stopping p at $x_n = 1/n$ (not to be confused with the necessary and sufficient condition for God_n to stop p at $x_n = 1/n$, which is what (2) provides, and which is now obtained by conjunction of this new H with the necessary and sufficient condition for God_n to actualize the possibility mentioned by this new H), God_n's capacity to stop p at $x_n = 1/n$ obviously being in such a case a necessary condition of H. As we now assume that God_n does not have this capacity, it is clear that not H is complied with (i.e. H is false). However, unlike what happened in section 2 of this chapter, now the conclusion of global interaction does not follow. The reason is that (3) may perfectly well be true (for each n), given that the mentioned incapacity of God_n is sufficient to make H false. The gods do not act globally here.

5. SHACKEL

Shackel (2005) has argued in favor of "the unsatisfiable pair diagnosis" applied to the different forms of the "Benardete Dichotomy":

"embedded in it is a pair of conditions which are jointly unsatisfiable:

1. Condition ANB: For all x in S, E at x iff E nowhere before x
2. The set S in condition ANB is an unbegun set" (p.401)

Sentence (1) (as ever, (1) and (2) refer, except where expressly indicated to the contrary, to the sentences so named in section 2 of this chapter) is definitely in accordance with this diagnosis: here the set S is the set of points $\{x = 1/n\}$ and Condition ANB is that for all x in S, a God stops the particle at x iff no God stops the particle before x (supposing that the set S is ordered in accordance with its natural ordering). However, I argued above that the correct form of the first version of Benardete's paradox considered in this article is (2), not (1), such that Condition ANB is not really modelized. Indeed, both the ANB condition and (1) and (2) take the form of universal sentences in which the quantifier affects a material equivalence. But in ANB and (1) this equivalence is of the form $A \leftrightarrow B$ in so far as (2) is the form $A \leftrightarrow (B\&C)$. I also believe that Shackel's position is inconsistent in this point. Let us suppose that $\forall n$, God_n controls the point $x_n = 1/n$ through the simple procedure of placing at that point a (sufficiently weighty) barrier, at an instant some time previous to the instant t=0. He appears to accept my thesis that in this case we have an example of global interaction where the particle interacts with the set of barriers without interacting with any one barrier ("This point makes sense when there are barriers in place", p. 403) but in such a case we are dealing with a form of the Benardete Dichotomy not analyzable by the unsatisfiable pair diagnosis! It would therefore seem that my idea of substituting (1) by (2) to represent the real situation is more in accordance with Shackel's own intuitions on this point than the unsatisfiable pair diagnosis.

Alternatively, let us suppose that $\forall n$, God_n controls the point $x_n = 1/n$ not by the simple procedure of placing a barrier there at some instant of time before the instant t=0 but rather by the "esoteric" procedure of creating a (previously non-existent) barrier there in the case that, for example, particle p has already passed point $x_{n+1} = 1/(n+1)$ on its way to x_n. This is the central point of Schackel's criticism as it effectively seems that, by applying my diagnosis, particle p will interact globally at x=0 with an inexistent set of barriers (in his words: "...but is more difficult to understand if the barriers are non-existent. In Benardete's Dichotomy, the man is stopped even though no god creates any barrier" (p. 403) and " ... the problem is being stopped by *non-existent* barriers" (p. 403)). When a colleague suggested this possibility to me for the first time, I thought it was of no great interest: if we open the door to peculiarities such as the creation of barriers out of nothing (and it must be from nothing, as otherwise particle p would interact globally with whatever it was existing previously) we should surely not be too scrupulous with regard to other similar oddities such as, for instance, the disappearance of particle p into nothing. However, after "Action without Interaction"(2005) and particularly the example of action without interaction put forward in this article (for which not even postulate PW(B) is necessary), my attitude changed on this point because now the disappearance of particle p into nothing (without interaction) is no longer an arbitrary anomaly but actually something acceptable (at least in the frame of a classical conception of the world, see note (6)). And since the creation from nothing is simply the temporal inversion of disappearance into nothing, an eminently acceptable supposition of temporal symmetry means this creation from nothing is no longer an arbitrary oddity. This way, my colleague's suggestion becomes well worth considering even when by accepting this we seem clearly to be heading away from the strict restrictions imposed by the principle of conservation of mass . Assuming then Shackel's intuition that particle p cannot interact with non-existent barriers and my result in the first part of this chapter, where p can never be in the region x>0 (the additional information that p will stop at x=0 or rebound there depended on the implicit supposition of pre-existent barriers) a new example of action without interaction follows: since (according to the model in the first part of this article under discussion now) p is moving for $\Delta<t<0$ in the region $x<0$ of the axis X according to the law $x=t$, and since it cannot reach the region $x>0$ despite not undergoing any interaction (its velocity must be 1 at every moment of its existence) p must disappear at $t=0$ so that it only exists in region $x<0$, and during $t<0$ instants, moving pursuant to law $x=t$. So the possibility of the gods creating barriers from nothing in the conditions mentioned leads to a new example of disappearance into nothing of particle p as a result of a global action of the gods not mediated by any barrier whatsoever or, therefore, by any interaction whatsoever (action without interaction).

Let us suppose finally that $\forall n$, God_n controls point $x_n= 1/n$ by the procedure of placing a barrier there in the case that particle p has already passed point $x_{n+1}= 1/(n+1)$ on its way to x_n, a barrier that for $t<0$ was at rest in $x^*_n = n+1$. As before, p cannot ever be in the region $x>0$ and, if we discard for now the option that in these conditions objects may suddenly appear or disappear, the only possible path is for p to rebound at $x=0$ or to stop there as a result of a global interaction at distance at $t=0$ with the set of barriers. In principle there is no problem with the idea of global interaction at distance: it would be propitiated by the capacity of the gods to move material objects (and therefore to make them interact by colliding with other objects) at velocities as great as one wishes. The problem arises with Newton's 3rd law (i.e. the "principle of action and reaction"): if the barriers interact globally on the particle by, for instance, stopping it, how does the particle interact on the barriers? Unlike the situation where the barriers were at rest at $x_n= 1/n$ for $t<0$ (in which case this interaction could take place by the induction of a self-excitation in the sense of Pérez Laraudogoitia (1996)) there now appears to be no non-arbitrary form of specifying that interaction, as at $t=0$ the distances between all the barriers (and also between them and the particle) are greater than a given finite non null number. This is, at bottom, Shackel's criticism of Hawthorne (2000) re-written in a language more in keeping with the content of this chapter concerning barrier-manipulating gods.[(8)] Hawthorne would argue that here the principle of action and reaction fails, i.e. in the terminology of note (8) what fails is the Change Principle ("This principle is mistaken, which I suggest is a big metaphysical surprise", Hawthorne (2000), p. 630), something that Shackel rejects ("Abandoning it ... Nothing wrong with that, but no interest in it either, and certainly no grounds on which to claim that the infinitude of [barriers in our case] caused the change", p. 410, "I suspect that changeless causation is an incoherent notion", p.408). I think, probably in agreement with Shackel, that one should not renounce the principle of action and reaction, at least in a classical conception of the world, but the alternative is not to adopt the unsatisfiable pair diagnosis and say, as Shackel does, that in situations like this we do "not allow any such executions, barriers ... to happen under a condition, but [make] them contradictory a priori" (p. 401-402). Indeed, since p is moving for $\Delta<t<0$ in the region $x<0$ of the axis X pursuant to law $x=t$ and given that it cannot reach the region $x>0$ despite the fact that it does not appear to experience any interaction whatsoever, the situation is essentially identical to the one discussed before in the case of the creation of barriers and so p must disappear at $t=0$[9], such that it only exists in the region $x<0$ and during instants $t<0$ (moving pursuant to law $x=t$). In both cases the implausibility of the interaction that would have to be introduced ad hoc means the action (which must undoubtedly be exercised on p) must be an action without interaction, i.e. disappearance at $t=0$[10]. Thus, Benardete's paradoxes also contribute, by means of the mechanisms of action without interaction they make possible, to highlight the problematic status of a principle of persistence in time that until now did not appear to be a subject for serious controversy. This contribution and the one introduced from the idea of global action show that these paradoxes imply a genuine metaphysical challenge. The unsatisfiable pair diagnosis is perhaps useful in elucidating logical paradoxes like Yablo's (1993) but to suggest that one can use it to clarify the various forms of Benardete Dichotomy is, in my opinion, to confuse logic with metaphysics.

Arguing in favour of his purely logical solution to the different forms of Benardete's dichotomy, (i.e. the unsatisfiable pair diagnosis), Shackel says that "in presenting the dichotomies no possibility has been specified" (p.413). Acknowledging also that there are principles implicit in the formulation of the dichotomies, he states: "Pérez Laraudogoitia […] seeks to strengthen the paradoxicality by rejecting the implicit principle whilst assuming that a philosophically significant result remains. I think applying the unsatisfiable pair diagnosis shows that rejecting the implicit principle is done at the cost of the significant result" (p.414). As my analysis thus far has shown, these implicit principles are in fact metaphysical in nature and amongst the philosophically significant results deriving from renouncing them (in the suitable contexts) is the possibility of global interactions and actions without interaction. For this reason, I think, unlike Shackel, that it is precisely the systematic application of the unsatisfiable pair diagnosis that (by being blind to all these possibilities) may only be performed at the cost of the truly significant results.

Someone might try to defend the option that Shackel's analysis is the correct one for Benardete-type situations by saying that when this analysis (basically his ANB condition) is not appropriate then it is not a proper Benardete-like puzzle. This would typically occur in cases of global action. Such a defence "by definition" has a number of problems or deficiencies:

a) One of the most typical examples of Benardete-like puzzles in the literature, the paradox of the gods (see for example Oppy, 2006, especially his section 3.6), would not in fact be a Benardete-like puzzle because it is possible to give a response to it in terms of global interaction, as we have seen in section 2 of this chapter.

b) This leaves open and unanswered the question of just when we can talk about global action and exclude the paradox. My proposal, to the contrary, provides indications in this respect: global action will only appear in circumstances in which the necessary and sufficient conditions of possibility are not complied with.

c) But, as we saw at the end of section 4, global action does not appear in absolutely all the circumstances in which the necessary and sufficient conditions of possibility are not complied with, so Shackel's analysis is also unsuitable in many cases where global action is absent. For another example in this direction, consider the following situation, in which we suppose that each God has the capacity to make the Empire State Building (ESB) collapse and we also suppose for simplicity's sake a given instant of time: God_n makes the ESB collapse on the surface of Mars in t if and only if (no God_{n+m} ($m \geq 1$) makes the ESB collapse on the surface of Mars in t and H**). Here H** provides the sufficient and necessary condition for the possibility that ESB collapses on the surface of Mars in t, a condition that certainly includes the ESB being on Mars in t. As this is not the case, H** is false and thus there is no paradox (contradiction): no God will do anything, there is no global action and Shackel's ANB condition is not met (it does not adequately represent the situation).

d) Implication-rich concepts that are fertile in multiple links are preferable in philosophical analysis. The concept of Benardete-like puzzle that I am now criticizing is not of this type, being in the last instance a purely formal standard contradiction. Benardete seems to have had a very different idea in mind, as reading Benardete 1964 makes clear. And the concept of Benardete-like puzzle I have used in this chapter (although I make no explicit attempt to specify it) is likewise richer: Benardete-type situations seem in many cases to be linked to interesting physical and/or metaphysical principles concerning reality (as described by a theory in the context of which the situations mentioned are analyzed).

6. On Angel

Shackel deserves praise for having recognized, in the situation described by Angel (2001), a model of the Benardete Dichotomy. In reference to this situation, let us designate for t_n the instant in which the pair of particles NP_n, PP_n collide (simultaneously) with M under the supposition that none of the particles NP_m, PP_m, ($m \neq n$) are present. The strictly decreasing succession of instants of time t_1, t_2, t_3, … converges there to a finite t_0. The model of Benardete's Dichotomy illustrated here according to Shackel is essentially the same as the one introduced previously by (1) and (1*). With the corresponding change in terminology:

(1**) The pair NP_n, PP_n collides with M in t_n if and only if no pair NP_{n+m}, PP_{n+m} ($m \geq 1$) collides with M at t_{n+m}

As should be clear from my arguments in this chapter, (1**) provides the necessary and sufficient conditions for the pair NP_n, PP_n to actualize the possibility of colliding with M in t_n ASSUMING THAT THIS POSSIBILITY IS ALREADY GUARANTEED. Therefore, (1**) is defective for the same reason as above: it takes no account of the necessary and sufficient condition for the possibility that the pair NP_n, PP_n collides with M in t_n. This condition is obviously that NP_n, PP_n is at null distance from M[(11)] at t_n. When it is added, we arrive at:

(2**) The pair NP_n, PP_n collides with M in t_n if and only if no pair NP_{n+m}, PP_{n+m} ($m \geq 1$) collides with M at t_{n+m} and the pair NP_n, PP_n is at null distance from M at t_n.

Analogously to the discussion in section 2 of this chapter, since (1**) leads to contradiction, it follows logically from (2**) that there are infinite values of n for which the pair NP_n, PP_n is not at null distance from M at t_n, i.e.

there are infinite values of n for which the pair NP_n, PP_n does not collide with M at t_n. But since $t_1, t_2, t_3, \ldots$ converges at t_0, if M were to continue at rest in some temporal interval $[t_0, t_0 + \varepsilon)$ (however small it might be) then there could only be at the most a finite number of values of n for which the pair NP_n, PP_n does not collide with M at t_n. In conclusion, M cannot be at rest in any interval $[t_0, t_0 + \varepsilon)$, $\varepsilon > 0$, and since, given the conditions of the problem, it is at rest at $(-\infty, t_0]$ it follows that it undergoes an interaction at $t = t_0$. Furthermore, as at $t = t_0$ all the pairs NP_n, PP_n are at a finite distance from M, it also follows that no such pair interacts at t_0 with M. In this way the set of pairs NP_n, PP_n interacts globally with M at t_0 without any pair interacting with M separately. Finally, as at $t = t_0$ the distance between the set of pairs NP_n, PP_n and M is also non null, it also follows that the interaction between the set of pairs NP_n, PP_n and M at t_0 is a non-contact global interaction. The interesting consequence of this, which corroborates the thesis I have maintained in the previous sections of this chapter, is once again clear: when we add to (1**) the necessary and sufficient condition for the possibility that the pair NP_n, PP_n collides with M at t_n, we arrive at (2**) and now the existence of a global interaction at distance at t_0 follows from here as a logical consequence[(12)]. This invalidates directly one of Shackel's central criticisms: " ... Angel simply helps himself to the assumption that a non-contact interaction happens at a convenient time to avoid the contradiction ..." "We are not being offered well-motivated constraints for the time and place of these non-contact interactions" (p. 412). The other central criticism has to do with Hawthorne's Change Principle and goes as follows: "... The existence of non-contact causal interactions would entail the falsity of the Change Principle, since we would have a case of changeless causation in which no change occurs to the set of raining particles but they cause a change in M. [It] is therefore objectionable for similar reasons to those I gave when discussing the Change Principle" (p. 412). I have already indirectly argued (see note (10)) that the Change Principle is false. Here I would like to point out, to conclude, that if the existence of non-contact causal interactions is objectionable "since we would have a case of changeless causation", then for the same reasons many cases of contact causal interaction (very different, therefore, from the model Angel proposes) would also be objectionable. Consider, for example, the frontal collision between two rigid spheres A and B. A has mass m and travels at velocity v towards B (which is at rest). B is formed by an infinity of concentric spherical layers C_0, C_1, C_2, C_3

… where C_i has mass m, internal radius i/(i + 1) and external radius (i + 1)/ (i + 2), so the mass of B is infinite. A standard result of the theory of collisions indicates that in this case, after the collision, A will rebound at velocity –v while B will remain at rest.[(13)] Obviously, B causes through contact a change in the velocity of A and this is a changeless causation in which no change occurs to the set of hollow spheres C_i but they cause a change in A. It is clear that the loser here is the Change Principle.

Notes

(1) As a particle world line is a subset of points of the spacetime, these postulates of continuity and connectedness may be united in a formulation in strictly topological terms of the following kind: "Particle world lines are path connected". For a definition of path connected see, for instance, Buskes and van Rooij (1997)

(2) Although I am simplifying slightly the description of the possible actions of the gods here, the simplification is innocuous for our purposes. Going into more detail, we may suppose that, for each n, God_n controls point $x_n = 1/n$ by means of a certain particle p_n identical to p (and, like p, a point particle) initially at rest at a distance $1/10^n$ from point x_n (but outside the X axis of p's movement). Now, if p has passed the mid point between $x_{n+1} = 1/(n+1)$ and $x_n = 1/n$ on its way to x_n and God_n has decided to stop it, he places p_n at rest at x_n before the instant $t_n = 1/n$ of p's arrival there. After the collision with p_n, p will be at rest at $x_n = 1/n$ and p_n will move to the right at the velocity unit that p initially had. Then, before p_n arrives at the mid point between $x_n = 1/n$ and $x_{n-1} = 1/(n-1)$, God_n will remove it from the X axis (it is not necessary to stop it to do this, it being possible by smoothly changing, for example, only the direction of the velocity of p_n, not its module). If p has passed the mid point between $x_{n+1} = 1/(n+1)$ and $x_n = 1/n$ on its way to x_n and God_n has decided not to stop it, then the god does not manipulate p_n so that p continues its inertial movement at velocity unit after passing point $x_n = 1/n$. In all this, we suppose that the velocity unit (initially possessed by p) is much lower than any of the upper bounds of velocity that, through the non-assumption of PW(B), might exist. This way there is no problem in God_n

inspecting the environs of the point x_n in time to block (if that is his purpose) the passage of p through it by means of a smooth movement and with a bounded velocity for p_n.

I shall not, of course, explain here how it is that God_n may "inspect" the environs of a point. In sections 2 and 3 of the chapter, I am not trying to construct models of Benardete's paradox IN classical particle dynamics but rather ones that are COMPATIBLE with classical particle dynamics (although, of course, without assuming any instance of PW(B)). This is also relevant with regard to another characteristic of my model: I admit that the collision between p and p_n is determinist and in it p transfers all its momentum to p_n. In fact, as we are dealing with point particles, this collision is indeterminist and the final state of rest of p is only one of an infinity of possible final states. So, the temporal evolution described above is compatible with dynamics, but not demonstrable in dynamics. In any case, this last limitation is easy to rectify: since, after the collision with p_n, p may remain at rest or be scattered it is a simple task to verify that the main argument in sections 2 and 3 of this chapter remains valid if we substitute "stops" for "stops or scatters" (as the necessary and sufficient conditions for possibility, which we shall use later in the chapter, do not change).

(3) It might appear at first sight that (2) does not correct the contradiction involved in (1): if $(\forall n)$ God_n stops p at point $x_n = 1/n$ should p actually arrive there, then it would seem that if no God_{n+m} $(m \geq 1)$ stops p, p will pass the point $x_{n+1} = 1/(n+1)$, in contradiction to what follows from (2). In fact, from the supposition that no God_{n+m} $(m \geq 1)$ stops p, it does not follow that p will pass point $x_{n+1} = 1/(n+1)$. To see this, let us suppose that, $(\forall n)$, God_n is simply a particle q_n identical to p and placed at rest at $x_n = 1/n$ before $t=0$. Now there are no gods with purposes, just particles. Obviously q_n stops p at point $x_n = 1/n$ should p actually arrive there (remember the problem is uni-dimensional and q_n is identical to p: elementary dynamics states that in the collision there will be a simple exchange of velocity). What is the result of this situation? Clearly, p cannot pass to the region $x > 0$, because for this to occur, it would have to be not stopped by an infinity of particles q_n and we know, to the contrary, that it would be stopped if it arrived at any point occupied by any of these particles. So p will be stopped at $x = 0$ or will rebound there once again towards the region $x < 0$. This implies that p interacts with the set of the

q_n without interacting with any particular one, and is an example of global interaction. There is no contradiction here: p never reaches the region $x > 0$ and so, from the fact that no q_{n+m} ($m \geq 1$) stops p, it does not follow that p will pass the point $x_{n+1} = 1/(n+1)$. If we now add the postulate that nothing interacts with p if it is not one of the particles q_n, (in the conditions of our model this means that p may interact at the most with one q_n) we obviously arrive at a contradiction. But all this does is to reduce to absurdity the supposition that nothing interacts with p if it is not one of the q_n, thereby recovering the idea of global interaction: p interacts with the set of q_n, without doing so with any one in particular. (For another simple dynamic model illustrating this same conclusion, in an even more "natural" way, see Pérez Laraudogoitia, 2005b, pp. 324, 325.) Translating the above into the language of the gods: although no God_{n+m} ($m \geq 1$) stops p, it does not follow that p will pass point $x_{n+1} = 1/(n+1)$; what follows is that p interacts globally with the set of gods (or the particles they control, if we use the model in note (2)) without doing so with any God_n in particular, stopping (or rebounding) at $x = 0$. As I said at the beginning of section 2: "only the gods … may affect p in one way or another". This they do by interacting globally (collectively) with p without interacting with p distributively (i.e. with no God_n interacting individually with p). Once again, the additional postulate that nothing interacts with p if it is not one of the gods (which means that the gods may only interact individually with p) leads to a contradiction and, once again, this contradiction simply reduces such a postulate to absurdity, thereby recovering the idea of global interaction. This is the idea that is automatically implied if one describes the situation by means of (2) instead of (1), as I defend in this chapter should be done.

(4) The conclusion of the sudden disappearance of p at $t=0$ could be avoided by supposing that p abandons its uniform movement some time before the instant $t=0$. If, however, its new kinematic law of motion $x=f(t)$ (with $f(t)$ a continuous function of t), were to lead to its having to arrive at $x=0$ at a certain instant t^* (i.e. $x=f(t)< 0$ for all $t< t^*$, and $0=f(t^*)$) we could not avoid the conclusion of its sudden disappearance at t^* (it would be enough to apply the argument given in the main text of the chapter, the sole difference being that we eliminate the universal of (4*) taking $t_n = t^*$ in place of $t_n = 0$). The only possibility of avoiding the sudden disappearance is, therefore, that p begins to decelerate before $t=0$ so that it never reaches the point $x=0$. But if we suppose

that p is not a charged particle, this deceleration would involve the introduction of a phantasmagorical interaction at distance by the gods (which we assume, as is also made clear in notes (2) and (3), interact by contact on the particle, should they interact on it), a completely "non-physical" interaction of repulsion. We are authorized to reject this alternative if, reasonably, we ONLY admit forms of interaction already accepted in classical physics in my models of Benardete's paradox.

(5) If two gods and only two gods want to do exactly the same thing at exactly the same time (namely, stopping p at $x = 0$ in $t = 0$), only the more powerful of the two will do it, this action obviously implying the annulment of the power of the other to do so. In the main text of the chapter I also implicitly assume that the power of any god to stop p (at $t = 0$ in $x = 0$) is only annulled by another, more powerful god stopping it. This supposition is necessary because otherwise, given that in the list God_1, God_2, God_3, ... no god is the most powerful, in principle it might be the case that none of them act on p thus leaving the particle to evolve freely for $t \geq 0$ as it did for $t<0$.

(6) The following observation is pertinent here. That the appearance and disappearance of particles is admissible in classical physics does not mean that, in such physics, material particles can appear or disappear more or less at whim depending on the circumstances. The principle of conservation of mass in the formulation of (Earman, 1986) says that:

C : particle world lines do not have beginning or end points

and

CC : mass is constant along a world line

In this particular context C has the following interpretation

C* : the particle World lines are maximally extended with continuity.

This allows us to understand how a particle may be non-eternal: for example when it is taken to the infinite in a finite time. Let us, however, consider a different situation in which a certain particle R is at rest at $x = 0$ for $t < 0$ and let us identify $x = 1$ as Q. Let $\forall n \geq 1$, the plan of action of god-n is as follows: whatever the point p of space is, provided that $p \neq Q$, if the particle R is at p at t = 1/(n + 1) then he will take it to occupy Q at t = 1/n and will make sure it stays at Q for t > 1/n (if the particle R is at Q in t = 1/(n + 1) , he will simply make sure it stays at Q for t > 1/(n + 1)). It is immediately clear (and this is how Pérez Laraudogoitia (2005) sees it) that, although the world line of R for t < 0 may be maximally extended with continuity for t > 0, this cannot be done in any way compatible with the restrictions on R imposed by the gods. This shows that C* must be weakened to take the following more qualified form (see for example Pérez Laraudogoitia (2009))

C** : the particle World lines are maximally extended with continuity in a way compatible with the restrictions on them

I have used the principle of continuity to explain in section 3 of the chapter (in a way compatible with C**) why p suddenly disappears in t = 0 but C** explains why the disappearance of p is not an available option in the scenario of Benardete's paradox analyzed in section 2.

(7) To clarify this point further, let us consider for example:

(A*) God_n bakes the cake P if and only if (no God_{n+m} ($m \geq 1$) bakes the cake P and particle p does not disappear suddenly at t=0)

Our unifying schema would entail replacing (A*) with

(B*) God_n bakes the cake P if and only if (no God_{n+m} ($m \geq 1$) bakes the cake P , the particle p does not suddenly disappear at t=0 and H*)

Assuming that H* is true we are led to (A*) i.e. to the conclusion that p disappears suddenly at t=0. However, this contradicts our background knowledge to the effect that the possible sudden disappearance of p has nothing to do with the gods baking cakes or not (in general, it has nothing to do with the performance or non-performance of operations which, if carried out, would not entail any interaction with p). This contradiction reveals that the new state of things posited is also impossible (to the extent that H* is true). Note that in this example the background knowledge plays an important role, like the one of the postulate that particles are eternal in the context of Benardete's paradox in section 3: but, just as renouncing the first is quite clearly nonsensical, renouncing the second means opening up an interesting path for exploration (indeed, classical physics does not assume that particles are eternal).

(8) Shackel's model (reminiscent of Hawthorne's, in turn reminiscent of Benardete's (1964)) involves a victim instead of a particle p and an infinite set of shooters (who, situated at a distance greater than k>0 from the victim, may shoot bullets at the victim successively at a velocity that grows progressively without limit) instead of an infinite set of barriers which, situated at a distance greater than k>0 from the particle, may be displaced successively by the gods at a velocity that grows progressively without limit. The important point that these two models have in common is that, in the case of the shooters "the boundary of the fusion of these bullets ... [is such that] ... there is no such 'limiting' boundary to hit the heart [of the victim] at time zero" (Shackel, p. 406) and, analogously, in the case of the gods, the boundary of the fusion of the barriers is such that there is no 'limiting' boundary to hit particle p at time zero. Finally, the analogue of Newton's 3rd law in my case is what Hawthorne calls the "Change Principle":

"If x is the fusion of y's and y's are individually capable only of producing effect e by undergoing change, then x cannot, (without the addition of some non-supervening causal power), produce effect e without undergoing change" (p. 630)

(9) If we suppose that in the present case material objects persist in time (as seems singularly plausible in light of the principle of conservation of mass for Shackel's model referred to in note (8)) then we have a contradiction. But this contradiction reflects the incompatibility of the underlying metaphysical suppositions (perhaps PW(B) is responsible for this, which would indicate an interesting, and unexpected incompatibility between PW(B) and the supposition of the persistence in time of material objects) and not, as the unsatisfiable pair diagnosis claims, an a priori impossibility linked to a certain logical form common to all the versions of the Benardete paradox.

(10) As in all examples of action without interaction, we have here a model of changeless causation, which refutes Shackel's suspicion that changeless causation is an incoherent notion.

(11) That A is in dynamic contact with B at t is necessary and sufficient condition for the possibility that A and B collide at t. However, dynamic contact between A and B at t reduces to contact *simpliciter* at t if, in the case

that its velocities at t^- were to remain constant from here on, A and B would interpenetrate after t.

(12) The existence of a global interaction at distance is a logical consequence of (2**), but (2**) presupposes that no pathological self-excitations that might spoil everything (e.g. self-excitations of the vacuum) actually take place!

(13) Let us suppose that two rigid spheres in one-dimensional movement of masses m_1 and m_2 and velocities v_1 and v_2 (in the referential of inertia S) collide frontally and elastically. Their velocities after the collision are $v'_1 = [(m_1-m_2) v_1 + 2 m_2 v_2] / (m_1+m_2)$ and $v'_2 = [(m_2- m_1) v_2 + 2 m_1 v_1] / (m_1+ m_2)$. At the limit where $m_2 \to \infty$ we are dealing with a collision between a particle of mass m_1 and another of infinite mass, in which case the final velocities will be $v'_1 = -v_1 + 2v_2$ and $v'_2 = v_2$. Transforming to an inertial system of reference S* that moves at velocity v_2 with regard to S we have a situation where, before the collision with a particle of infinite mass at rest, the colliding particle moves at velocity $v_1 - v_2$. After the collision, the particle of infinite mass continues at rest while the colliding particle acquires the new velocity $-v_1 + v_2 = - (v_1 - v_2)$, i.e. its velocity changes sign after the collision (see for example Spiegel (1967), chapter 8, problems 8.7 and 8.24). As we know, the process of passing to the limit can be very problematic, in that it can lead to conflicting results. But this is not the case here, as no conflicting result arises from its application. The fundamental principle of the conservation of linear momentum would seem to be violated, because in system S* the total momentum before the collision is $m_1(v_1 - v_2)$ and afterwards, $-m_1(v_1 - v_2)$; however, no such violation really occurs, precisely as a consequence of the conservation of linear momentum being (let us acknowledge the fact) a fundamental law of nature. It follows from this, under Galileo's principle of relativity, that if the law of the conservation of momentum is applicable to process Q in the inertial system of reference R, it will be likewise applicable in any other inertial system of reference. But the law of conservation of momentum is not applicable to our example of a collision in referential S (in S the momentum is infinite before and after the collision, but the equality $\infty = \infty$ is meaningless), which means that it is not applicable in S* either. So although in S* the momentum before the collision is different from the momentum afterwards, no fundamental law of the conservation of momentum is violated. In fact, the pervious passage to the limit $m_2 \to \infty$ is as innocuous as the similar

passage made when one solves the problem of two bodies to which the model of the hydrogen atom with nucleus of infinite mass leads (which is precisely what Bohr did in his first atom model, Petruccioli (1993)).

References

Angel, L. 2001. A Physical Model of Zeno's Dichotomy. *The British Journal for the Philosophy of Science*. 52: 347-358

Benardete, J. 1964. Infinity: An Essay in Metaphysics. Clarendon. Oxford

Buskes, G. and van Rooij, A. 1997. Topological Spaces. Springer. New York

Earman, J. 1986. A Primer on Determinism. Reidel. Dordrecht

Oppy, G. 2006. *Philosophical Perspectives on Infinity*. Cambridge University Press. Cambridge

Hawthorne, J. 2000. Before Effect and Zeno Causality. *Noûs*. 34: 622-633

Pérez Laraudogoitia, J. 1996. A Beautiful Supertask. *Mind*. 105: 81-83

Pérez Laraudogoitia, J. 2003. A Variant of Benardete's Paradox. *Analysis*. 63: 124-131

Pérez Laraudogoitia, J. 2005. Action without Interaction. *Analysis*. 65: 140-143

Pérez Laraudogoitia, J. 2005b. An Interesting Fallacy Concerning Dynamical Supertasks. *The British Journal for the Philosophy of Science*. 56: 321-334

Pérez Laraudogoitia, J. 2009. Physical Action Without Interaction. *Erkenntnis*. 70: 365-377

Petruccioli, S. 1993. Atoms, Metaphors and Paradoxes. Cambridge University Press. Cambridge

Shackel, N. 2005. The form of the Benardete Dichotomy. *The British Journal for the Philosophy of Science*. 56: 397-417

Spiegel, M.R. 1967. Theoretical Mechanics. McGraw-Hill Book Co., Inc. New York

Yablo, S. 1993. Paradox without Self-Reference. *Analysis*. 53: 251-252

In: Philosophy of Science
Editors: S. Pintuck and C. Reynolds
ISBN: 978-1-62100-276-5

Chapter 2

POINCARÉ'S ENIGMATIC RELATIONSHIP WITH LOGIC AND THE INFINITE: NEW INSIGHTS INTO THE RELEVANCE OF INTUITION TO HIS ANTI-LOGICIST PROGRAMME

Margaret MacDougall
University of Edinburgh Medical School, UK

ABSTRACT

Traditionally, Henri Poincaré (1845 - 1912) has been recognized as a pioneer in the discovery of predicative set theory. Less attention, however, has been attributed to the importance of intuition to his interpretation of what it means to exist in the domain of pure mathematics. This imbalance is hard to justify in terms of Poincaré's own philosophical writings, as his contentions with his contemporaries – most notably, the logicists – are typified by appeals to the indispensability of intuition as a foundation for mathematics. The value of predicativism as a means of circumventing the set-theoretic paradoxes and indeed, of rendering mathematical definitions and proofs more transparent is well-recognized. Nevertheless, the associated systems are also noted for their restrictive nature in terms of proof-theoretic strength. It is important, therefore, to take a fresh look at Poincaré's philosophical writings to establish how the divergence between his foundational arguments and his reputation as a predicativist may have emerged. In this chapter, I shall offer new insight into the relevance of intuition to Poincaré's criteria for set existence and in particular, his requirement that infinite sets should be surveyable. On this basis, I shall argue that Poincaréan set theory ought to be more liberal than has traditionally been assumed. In exploring the above criteria, I shall also revisit Poincaré's unexplained affinity towards classical logic. By appealing to the influence of Aristotle's writings on Poincaré's interpretation of logic, I shall show that his ambiguous reference to the intuition of pure number as that of pure logical forms invites further exploration. Thus, I shall argue that Poincaré's treatment of the principles of Non-contradiction and Excluded Middle as necessarily true has its roots in logical intuition – a variety of intuition which he neglects at the cost of his own anti-logicist stance. These findings serve as an extension of my recent work on the nature of mathematical intuition. As such, they provide some essential groundwork for developing an alternative set theory with the potential to create a more successful synergy between Poincaré's philosophy and practice in pure mathematics.

1. The Intuitive Basis for Poincare's Notion of Set

Poincaré recognizes sets as entities which we need to *construct* rather than *discover*. In so doing, he requires that during the construction of a set, as new members are added, set membership should be conservative in terms of admissibility of previous members. Thus, any infinite set which can be deemed to exist should be in principle constructible - that is, it should be possible to add new members indefinitely without ever needing to eliminate set members previously admitted. It must also be dependent on an immutable classification – a "predicative classification" – of objects into membership and non-membership of that set. Furthermore, Poincaré recognizes the creation of each set member as occurring in an immutable order represented by a law of correspondence between the set members and N- $\{0\}$, where N is the set of natural numbers. The immutability of set membership is thus carried over to this law, referred to as such by Poincaré as a *predicative law of correspondence*. [Poincaré 1963, pp. 49 – 50, 62] By contrast, any change in the above classification would require that the law of correspondence be re-defined.

In keeping with the above, Poincaré identifies the following two criteria as fundamental to the recognition of a genuine set [Poincaré 1909b]:

C1) The associated predicate must be definable in a finite number of words

and

C2) There must exist a predicative law of correspondence between the members of ***N*** - {0} and the members of the given set.

Here, Poincaré is adopting a form of *finitism* – a philosophical position which in its many guises [Tiles 1989] typically adheres to the view that our interpretation of the infinite in mathematics must be informed by the limitations of our mental capacities. Thus, in particular, Poincaré's criteria for identifying a genuine set give precedence to the requirement of being able to survey the set as a construction - that is, of recognizing that construction as consisting in either a finite or a potentially infinite number of steps. For Poincaré, in the case of an infinite construction, essential to such an activity is the knowledge that no earlier step in the construction will require to be reversed on the basis of any subsequent step. Thus, when reflecting on what he recognizes as pure mathematics, he defends the general doctrine that mathematical activity requires one to

> "[build] up constructions more and more complicated." [Poincaré 1946b, p. 382]

A necessary and sufficient condition for a set ***C*** to be constructible by means of an immutable classification is that ***C*** be *countable* in the sense specified by Bishop [Bishop 1967] that there exists a surjection from ***N***- {0} to ***C***. In the case of infinite sets, this condition amounts to denumerability and such sets are often referred to as countably infinite. In the case of finite sets, in Poincaréan terms, their surveyability is established *ipso facto* by their *finiteness*.

Poincaré illustrates what it means for a set to be uncountable in his treatment of the paradoxical set, ***E*** – *the set of all numbers definable in a finite number of words* – associated with Richard's paradox. He argues that for this set, Richard's proposed way out of the paradox, through redefining set membership indefinitely, leads to violation of C2 above. [Poincaré 1909b] As Poincaré explains, the only foreseeable way of rendering classification of set membership immutable in this case would be to assume that ***E*** already exists. But, as he goes on to explain, for this one would require to recognize ***E*** as existing objectively as an actually infinite set, which is contrary to Poincaré's treatment of infinite sets as indefinitely extendible progressions. In fact, for Poincaré, it is ultimately the assumption of the existence of the actual infinite which makes provision for the emergence of the *petitio principii* responsible for Richard's paradox. [Poincaré 1946b, p. 480] More generally, Poincaré treats the requirement to assume the existence of the actual infinite as a diagnostic criterion for the existence of an improperly defined set and such a requirement is symptomatic of definitions for uncountable sets.

With regards to the need forC1 and C2 as criteria for sethood, it is informative to consider Poincaré's dictum,

> 'What does the word exist mean in mathematics? It means . . . to be free from contradiction.' [Poincaré 1946b, p. 474]

In a rejoinder to Poincaré, Couturat argues that this dictum is a rather exacting one, emphasizing that

> "The presence of a contradiction can certainly be proved, but the absence of every contradiction is, like the innocence of an accused person, a negative fact which cannot be proved directly." [Couturat 1912, p. 513]

So as to provide an adequate response to this rejoinder,Poincaré is duty bound to provide criteria for sethood which rule out the possibility that their defining predicates should provide the basis for a contradiction. It is likely, therefore, that Poincaré's C1 and C2are intended to serve this purpose.

However, as will be explained more fully later, Poincaré also has a concern for the future of sets under Russell's emerging theory of *propositional functions* whereby set existence is treated as at most apparent. Thus, he chooses to nail his colours to the mast by presenting "absence of contradiction" as the only "directing rule" for deciding which definitions have a class as an extension. [Poincaré 1946b, p. 479]

As Poincaré recognizes the intuition of pure number as"the fountainhead of arithmetic and unique as a source of certaintyand rigour" [MacDougall 2010] and identifies *the principle of mathematical induction* (*PMI*) as representative of "reasoning par excellence" [Poincaré 1946b, p. 452], it wouldnot be too surprising if he regarded the "the principle of induction" when viewed "as a synthetic judgement a priori" [Poincaré 1946b, p. 475] as his guarantor of consistency. If Poincaré is to adhere to this foundational position regarding pure mathematics there ought to be evidence to suggest that at least one of C1and C2are grounded on *PMI*.

Notice therefore that Poincaré's interpretation of the construction of an infinite set as a sequential and immutable process is itself dependent on *PMI*, where the predicate to be satisfied at any given stage is *preservation of the order of the elements constructed up until now*. This would suggest that Poincaré's notion of construction is grounded on *PMI* and *ipso facto*, on the intuition of pure number. [MacDougall 2010]

It is not evident from Poincaré's writings that he recognized this connection between sethood and intuition; otherwise, one would have expected him to have exploited it in his critique of the logicists. Nevertheless, he would have been supportive of a programme designed to capitalize on such a connection. This is particularly clear from his recommendation that,

> "Criteria are needed to decide whether a definition is too complex or too extended, and these criteria can only be justified by an appeal to intuition." [Poincaré 1946b, p. 480]

Thus, in so far as sets themselves are defined by predicate definitions and criteria for definitions are in turn criteria for set existence, Poincaré's brand of finitism may also be read as requiring that sets are *intuitive*. Further, notice that according to C1 and C2, *N* is itself countable, and can therefore be included among the sets which have an intuitive basis for existence.

We can go further back than the immediate writings of Poincaré, however, to identify a close connection between his C2 and dependency on intuition. On considering the Greek notion of *Nous*, with its close connections to that of *intuition*, [Smith 2007] it is particularly interesting to note that this faculty has been recognized as consisting in understanding the "essential immutable nature of things". [Jones 1995] In his affinity to intuition as a foundation for mathematics, perhaps Poincaré was automatically endorsing this property in his specification of C2, above.

Poincaré's identification of predicative classifications as fundamental to set existence is, however, also entirely in keeping with his contemporary Frege's requirement that sets be *decidable* objects, where for an arbitrary set $\hat{y}A(y)$ to be decidable, the formula

$$\forall x(A(x) \vee \neg A(x))$$

must be true, which is to say the set predicate should satisfy (*the Principle of*) *Excluded Middle* (*PEM*). In turn, the decidability of sets is essential to their status as Fregean proper (or, bona fide) objects. [Frege 1964, p. 128]. It is implicit from Poincaré's treatment of sets in this manner that these two philosophers also have in common the assumption that sets exist as extensions of predicates – that is, that their members are posterior to their defining predicates. However, in requiring that decidability be determined in an irreversible stepwise manner, Poincaré is pursuing a constructive approach to set existence not evident from Frege's philosophy of pure mathematics. Thus Poincaré's association of sets with proper (or, immutable) classifications may be alternatively stated as a requirement that all his sets be constructed so as serve as a constructive equivalent of Fregean proper objects.

It is important to note that there is a distinction, however, between those sets excluded by Poincaré's intuitively grounded notion of uncountable set and those excluded byhis notion of *impredicative definition* as a definition involving quantification over a totality which A) includes the object being defined or B) whose existence depends on the object being defined.[Zermelo 1908, p. 190]The above interpretation of *impredicative definition* is implicit from Poincaré's philosophical writings, especially those involving his contentions with the logicists.

Philosophers of the latter camp are intent on reducing mathematics to logic, so that at least in Frege and Russell's case, mathematical, including arithmetic and pure geometric, definitions and propositions, are thought of as essentially derivable from a restricted set of logical statements. Further, a mathematical statement is true on analytic grounds – that is "owing to the meaning of the concepts occurring in it." [Gödel 1944] Based on the latter view in particular, numerous attempts have been made by Poincaré's contemporaries to present the *PMI* as falling out of an analysis ofthe concept of natural number rather than, as Poincaréwould have it, a principle which we arrive at directly through the influence of the intuition of pure number.In Poincaré's view, all such attempts involve a vicious circle as all attempts to define natural number use words which themselves rely on the concept of number. Thus, analysis of concepts cannot serve as an adequate foundational tool for the affirmation of *PMI* as a valid principle of pure mathematics.. [Poincaré 1946b].

It is clear from a more careful consideration of Couturat's rejoinder to Poincaré [Couturat 1912] that Poincaré takes part b of the definition of impredicativity above to extremes through failing to distinguish between meta-mathematical language and mathematics proper. Nevertheless, the meaningfulness of this form of impredicativity in pure mathematics itself is evident from Parsons's representation of the *PMI* as a rule of inference according to the following definition, involving the free predicate variable *A*and the predicate constant *N* for *is a natural number* :

$$\frac{A(0) \qquad N(\mathrm{a}) \rightarrow [A(\mathrm{a}) \rightarrow A(S\mathrm{a})] \qquad N(t)}{A(t)}$$

As is implicit from Parsons's explanation, the range of the free variable *A* includes second order predicates and hence predicates involving the predicate *N*. [Parsons 1991]

The latter illustration of an impredicativity of type B explicitly involves the entire set being defined. However, an impredicativity may also occur in the definition of a number in terms of a set where the number is itself a member of that set, leading to a predicativity of type A, above. Such is the case when defining the least upper bound or greatest lower bound of a previously defined infinite set of numbers within a proof with a view to making further inferences (resulting in quantification over *all* upper and lower bounds, respectively). [Zermelo 1908] In fact, Argand and Cauchy's proof of the fundamental theorem of algebra involves an impredicativity of this very sort through the assumption of the existence of a lower bound to the values of a chosen real-valued function [Chrystal 1904]. As Zermelo highlights, it is for this reason that Poincaré is unable to entertain this theorem as meaningful. [Heinzmann 2006; Zermelo 1908] Regrettably, Poincaré fails to provide even a restricted form of this theorem and a corresponding proof which are faithful to his own philosophy of pure mathematics.

In making these observations, it is important, however, to bear in mind that the term*predicative* originates in Poincaré's era with Russell's idea of specifying which definitions have a set as an extension and as such are true predicates, and that this interpretation was at the heart of Poincaré's usage of the term 'predicative' in relation to immutable classifications. The fact that the notion of *impredicative set* has traditionally been associated with Poincaré is not rooted in his remarks on intuition but rather, in his contentions with Russell over the logical antinomies and in particular, Russell's proposal that a logical principle, the Vicious Circle Principle (VCP), could provide a solution to these antinomies. Russell's conviction that the circularity inherent in impredicative definitions and the associated logical paradoxes can be fully addressed by logic undergirds his appeal to the VCP. Various versions of the VCP exist depending on whether the impredicative definition is thought of as a predicate or propositional function and on what is understood as an impredicativity, where in the latter case, this can lead to controversy over what is meant by the true VCP. However, for simplicity here, it suffices to note that Poincaré is interested in a VCP which denies the existence of infinite sets for which impredicative definitions exist in the sense of A or B, above.

In his contentions with Russell over whether the VCP, as a logical principle, confirms the place of logic in solving the logical paradoxes, Poincaré fails to tease apart the notions of countability and predicativity or to fully delineate the intuitive basis for the former of these notions. Thus, at heart, Poincaré's writings have a much stronger intuitive basis for development of a semi-constructive set-theoretic system than his reputation as a forefather of predicative set theory and some interpretations of his maxim, "Avoid non-predicative classifications and definitions" [Poincaré 1963, p. 63] may suggest. Moreover, based on his failure to distinguish between the *countable* and the *impredicative*, the formulations of mathematics that have been assumed as a product of Poincaré's debates with Russell are more restrictive that would otherwise have been necessary purely through appeal to intuition in the above senses. Given that Poincaré is more taken up with refuting Russell's attempts at logical reductionism than exploring the nature of the sets that he himself is proposing, it is hardly surprising that he does not have examples of impredicative countable infinite sets at his fingertips. On the other hand, within a universe where all sets are countable, and therefore where the power set of **N** is absent, such examples are remarkably challenging to construct and tend to rely on the theory of countable ordinals, a theory which is not fully recognized within Poincaré's era.

Generally speaking, ordinals may be understood as the order types of well-ordered sets used to classify the order structures of the elements of these sets. Each ordinal explicitly recognized as such within a given formal system is canonically identified with the set of ordinals which precedes it, which is itself a well-ordered set. Thus, it makes sense to talk about countability of individual ordinals. Any countable ordinal including and beyond Γ_0, the well-ordered set of all predicative ordinals [Feferman 1964, p. 13],can be taken as a counterexample to the statement, *predicativity is a necessary condition for countability*. These particularcountable ordinals are those for which the existence of a well-ordering can only be proved impredicatively within a formal system. Their existencehas important implications in terms of making suitable inferences about the type of formal system which ought to emerge from treating intuition, as Poincaré intends, as the foundation for existence in mathematics. In particular, the least ordinal within a given formal system for which transfinite induction, and hence a well-ordering, is unprovable is typically recognized as a measure of the proof-theoretic strength of that system. Thus, it may be argued that insistence on predicativity rather than countability could lead to a weaker and hence a mathematically less productive system than that which ought to have emerged from Poincaré's account of the foundations of pure mathematics.

A possible way forward in developing countable set theory can be found through consideration of Holmes's proposal for Pocket Set Theory,a theory so named because of the smallness of the domain of sets it encompasses.On referring to infinite collections, Pocket Set Theory allows for the distinction to be made between *sets* which are countably infinite and proper classes, which have the same cardinality, $\aleph_1$, as the mathematical continuum and, as such, are uncountable. [Holmes 2006; 2007] In excluding proper classes, including their corresponding ordinal ω_1 (the first uncountable ordinal), from set-hood Holmes's proposal is in keeping with the requirement that a set be surveyable in Poincaré's terms. However, considerable work could be done in the future to compare the proof-theoretic strength of a fully formalized version of Pocket set theory with that of other formal systems. Of particular interest here would be consideration of the theorems provable within such a system by comparison with Zermelo set theory and its variants. Poincaré sees Zermelo's axiomatization of set theory as merely the product of a hacking procedure applied to the true notion of set in order to circumvent the set-theoretic paradoxes. [Poincaré 1963] Correspondingly, there is a clear need to replace Zermelo's "arbitrary decrees" by axioms which have a more evident foundation in intuition.

Russell's own VCP undergirds the interpretation of propositional functions (and apparently sets) in hierarchical form by stipulating as a proposed solution to the logical paradoxes that *no propositional function involving quantification over all of the argument range for a given variable includes the propositional function itself.* [Russell 1993] Russell presents the theory of ramified types, RT, [Russell 1967] as a stratified formal system which conforms to his VCP in terms of the ranges, or, logical types, of the free and bound variables belonging to propositional functions within the different strata of the hierarchy. As such, this theory is designed to circumvent the logical antinomies, including the famous *Russell paradox*, associated with the paradoxical *Russell set– the set of all sets which do not belong to themselves.*

[Tiles 1989] RT serves an additional purpose, however, namely "the reduction of propositions nominally about classes to propositions about their defining functions", which is undeniably a bone of contention for Poincaré. The endpoint which Russell has in view in terms of his logicist program is expressed as follows:

> 'The absence of all mention of particular things or properties in logic or pure mathematics is a necessary result of the fact that this study is, as we say, "purely formal".' [Russell 1993, p. 198]

While it is not my intention to explore RT or indeed, its contentious foundational status, here in detail, it is of particular relevance to consider the notion of propositional function assumed within this system and later within the more comprehensive deductive system contained within Principia Mathematica (PM)[Whitehead and Russell 1925], as this is essential to the arguments which follow on the nature of logical forms and hence of logical intuition.

2. Propositional Functions and Their Relevance to Poincare's Philosophy of Mathematics

A propositional function is a function expressed as a sentence involving one or more variables. For the sentence to take on a definite meaning and be judged as true or false, each of these variables requires to be replaced by the name of a definite entity. Any sentence so formed is a proposition.[1] Prior to this stage, the propositional function is undetermined. Alternatively put, it is "a mere schema, a mere shell, an empty receptacle for meaning, not something already significant" [Russell 1993] or "unsaturated" [Frege 1964]. As Russell,

[1]The view, implicit from Russell's writings, that any proposition is capable of being judged as true or false is one which ought to be carefully weighed up in the light of Gödel's ground-breaking incompleteness proof of 1931. This proof demonstrated effectively that in every formal system of arithmetic, including that found within Whitehead and Russell's deductive system in Principia Mathematica, there exist propositions which can neither be proved nor disproved within that system. Gödel, K. (1962). On Formally Undecidable Propositions of Principia Mathematica and Related Systems (B. Meltzer, Transl. with introduction by R.B. Braithwaite). (Edinburgh: Oliver and Boyd (First published 1931)). Additionally, there are many problems of modern mathematics for which no current solution exists. In such cases, propositions stating the existence of a solution for any one such problem are strictly speaking, neither true nor false. If, as many constructivists do, we reduce mathematical truth to provability, then it is incorrect to assume that PEM holds for all propositions.

indicates, "[a]ny mathematical equation" involving at least one undetermined variable is therefore a propositional function, as is any sentence of a formal logical system, including Aristotelian logic. [Russell 1993, p. 157] More generally, propositional functions could be regarded as sentences formed by abstraction from propositions bearing the same structure. This idea can be illustrated by means of the following three examples.

i. The propositional functions

***x* is president of the USA** (1)

and

A(x) (2)

are incomplete versions of the false proposition

Donald Cameron is president of the USA

formed by replacing 'Donald Cameron' by the subject variable '*x*' in the case of (1) and in addition 'is president of the USA' by the predicate variable '*A*' in the case of (2).

ii. The propositional functions

x* is less than *y (3)

and

xRy (4)

are incomplete versions of the true proposition

The average life expectancy of domestic cats in the UK is less than the average life expectancy of men in the UK.

formed by replacing 'average life expectancy of domestic cats in the UK' by '*x*' and 'average life expectancy of men in the UK' by 'y' in the case of (3) and in addition, replacing 'is less than' by '*R*' in the case of (4).

iii. The propositional functions

x is an integer and y is greater than x (5)
and

A(x)* and *yRx (6)

are incomplete versions of the true proposition

4 is an integer and 4.5 is greater than 4

formed by replacing '4' and '4.5' by 'x' and 'y', respectively in the case of (5) and in addition, replacing 'is an integer' by '*A*' and 'is greater than' by '*R*' in the case of (6).

In each of cases i to iii, infinitely many alternative true or false propositions could have been provided to form the same propositional functions. However, more importantly here, it is also clear that each pair of propositional functions themselves demonstrate varying degrees of incompleteness (with the second propositional function being more incomplete or abstract than the first). We shall return to the latter property later in connection with the notion of logical form.

In understanding propositional functions better, it is also helpful to note that there exist special cases where all of their instances are true. A few illustrations are provided below.

Within the domain of 19th Century classical logic, Aristotelian logic, in its original sense, [£ukasiewicz 1957] is comprised of propositional functions for which all corresponding propos*itions are true.* For example, consider the syllogism,

If a is predicated of all b
and b is predicated of all c (7)
then a is predicated of all c.

Within this syllogism, the notation is such that the interpretation of what the variables *a*, *b* or c signify is dependent on whether they lie to the left or the right of *is predicated of*, with variables to the left denoting predicates and variables to the right denoting subjects.

The syllogism itself constitutes a propositional function which is true for all predicate values of *a*, all subject values of *c* and all predicate and subject values of *b*. More generally, this property can be attributed to all the primitive or, foundational, principles which define any formal system of classical logic.

It is clear that in 1918, at least, Russell regards propositional functions from a logical point of view and sees them as a means of treating reference to sets[2] as a mere *façon de parler*,to the end that sets can no longer be regarded as objects but assume the status of "logical fictions". [Russell 1993]

This is perhaps best explained here at the simplest level where we avoid the philosophical complexities which might arise if we considered propositional functions containing variables which themselves range over propositional functions. Instead, taking the arguments of propositional functions to be more primitive than propositional functions, each propositional function in which any existing relational or predicate variables are instantiated is understood to determine a set, namely the set of values of the arguments for which the function is true – that is, the *extension* of the function or *argumentrange* for the function. For example, in illustration ii, above, if we allow the argument ranges of *x* and *y* to be all living creatures, one can imagine an indefinite number of alternative possible substitutions to the pair *<average life expectancy ofdomestic cats in the UK, average life expectancy ofmen in the UK>* which would still render the resulting proposition true. Such possibilities constitute the members of that set which is the extension of the propositional function (3), with relational constant 'is less than'. More significantly, here, however, a one-to-many mapping is understood to exist from sets to propositional functions. Precisely, every set can be defined in terms of a non-unique propositional function which is true for all members of the set but false otherwise. For any given set, Russell refers to all propositional functions to which the set can be mapped in this way as *formally equivalent.* Russell then claims that since no set can be mapped to a unique propositional function or reduced simply to its members (so as to be regarded as an "heap"), any set may just as well be regarded as a symbol. This argument is hard to swallow, however, unless, perhaps, propositional functions are assumed to be more foundational than sets. In fact, in a number of instances Russell seems to prefer to take the position of the agnostic regarding sets and indeed predicates and propositions [Russell 1984, p. 100, 110], suggesting in each of these three cases that the matter of their existence is unimportant.

---

[2] In Russell's terminology, the term *class* rather than *set* is assumed, but the term *set* has been used consistently throughout this chapter to avoid departure from the central arguments concerning the intuitive status of sets.

3. Propositional Functions and Logical Forms

It is clear, at least from some of his writings, that Russell treats logical forms as very closely tied to propositional functions, as is implicit from the following definition and illustration:

> 'The "form" of a proposition is that, in it, that remains unchanged when every constituent of the proposition is replaced by another.
>
> 'Thus "Socrates is earlier than Aristotle" has the very same form as "Napoleon is greater than Wellington," though every constituent of the two propositions is different." [Russell 1993, p. 199]

As such, logical forms express what is "logically possible" without the guarantee that their instances are actually true. [Russell 1984, p. 111, p. 130]

Furthermore, Russell also wants to make clear that the symbolic expression of a given logical form, which uses onlyapparent predicate, relational or subject variables (that is, variables representing nothing in particular) and which as such contains "no constituents" (or, real content), is not to be confused with the logical form itself. This intention is particularly transparent in the following remark:

> "A *form* is not a mere symbol: a symbol composed entirely of variables symbolizes a form, but is not a form." [Russell 1992]

Later, Russell illustrates this idea in a number of ways, especially with reference to the forms of all *subject-predicates* and *dual complexes*. These he regards as the facts that *something has some predicate* and *something has some relation to something*, respectively, while the corresponding symbolic expressions of these forms might be "$A(x)$" and "xRy" where A ranges over predicates and R ranges over binary relations. Thus, in particular, the above logical forms may be conveniently associated with (2), (4) and (6) in our consideration of propositional functions.

Taking the above deliberations into consideration, I think it would be more precise to say that based on a substantial contribution from Russell's work, *logical forms may be regarded as those entities which when represented in symbolic form constitute propositional functions with no complete variables.*

But what sort of entities are they in an ontological sense? A useful starting point in answering this question is to note that while referring to a process of abstraction from a proposition containing actual entities (e.g. Socrates precedes Plato) to the symbolic representation of the form within that proposition (*xRy*) and finally to the form itself [Russell 1984, pp. 98, 104, 114, 129], Russell also recognizes logical forms as *structures* and as *objects*. Russell further specifies that logical forms, while being logically (but not psychologically) simpler than the propositions which serve as their instances, or, values [Russell 1992], are at the same time "genuine" objects, on the grounds that we have acquaintance with them. [Russell 1984, p. 129] As we shall see shortly, this ontological position can be made more precise through the interpretation of logical forms as *objects of intuition*.

In referring to rules of deduction as particular types of proposition which have the dual role of serving as premises and providing consequences of premises, Russell distinguishes between the subject as inessential, and the form as essential, to its *validity*. [Russell 1911]

These clear distinctions between logical forms and subjects, both in terms of their ontological status andtheir significance, amounts to logical forms being given a more definite form of existence, including objecthood, with subjects being explained away as purely symbolic or nominal.

As to the origin of logical forms, some of Russell's views in this respect are captured by means of the following observations:

> "Mathematics is ... wholly composed of purely formal propositions... It is remarkable that we have the power of knowing such propositions." [Russell 1911]

> "[A]cquaintance with logical form ... is a primitive constituent of our experience..." [Russell 1984, p. 99]

and

> "Logical propositions are such as can be known *a priori*, without study of the actual world." [Russell 1993, p. 204]

These remarks are of fundamental relevance to any foundational argument which Poincaré might have wished to consider in defence of his notion of an *intuition of pure logical forms*. In particular, the latter variety of intuitiondemands a fuller treatment than that afforded in his philosophical writings on pure mathematics. This need is reinforced on considering remarks made by Couturat and Wittgenstein.

In the former case, on considering Poincaré's contrast between "the logical and the intuitive", Couturat complains that,

> "the vague conception of *intuition* is out of place as a weapon against the logicians, especially when the intuition spoken of is not specified." [Couturat 1912, p. 497]

Wittgenstein characterizes *logic* in his Notebooks of 1914 – 1916 [von Wright and Anscombe 1961] as the domain responsible for informing us as to what is possible in terms of "the [*a priori*] logical form of thought, language and the world", although not for the propositions which arise from these forms [Summerfield 1990]. While Wittgenstein is not a proponent of the logicist doctrine that mathematics can be reduced to logic, his above treatment of logic begs the question as to the mechanism through which logic acquires the above foundational status.

While criticizing Kant for his idealism, whereby logical truths are mind dependent, Russell stresses the objectivity of *a priori* truths, by which he intends "self-evident" truths, which can be "known without demonstration". [Russell 1984a] It is important to emphasize here, however, that in so doing Russell need not have dispensed with the role of intuition in equipping us with knowledge of these truths.

His recognition of a close relationship between logic and the physical world is evident from his remark that,

> "logic is concerned with the real world just as truly as zoology, though with its more abstract and general features." [Russell 1993, p. 169]

Clearly, this remark invites some justification on epistemological grounds. The fact that intuition might have a role to play here is implicit from Gödel's remarks on Russell below:

> 'He compares the axioms of logic and mathematics with the laws of nature and logical evidence with sense perception, so that the axioms need not necessarily be evident in themselves, but rather their justification lies (exactly as in physics) in the fact that they make it possible for these "sense perceptions" to be deduced; *which of course would not exclude that they also have a kind of intrinsic plausibility similar to that in physics.*' [Gödel 1944, p.127, my emphasis]

While, on the one hand, logical truths may be suited to our understanding of the material world, it is also legitimate to recognize the necessity of intuition in equipping us with the capacity to view the world the way we do. In fact, in one of his relatively few remarks on intuition, Russell himself almost moves in this direction by referring to "acquaintance with logical form" as 'that which may perhaps be called "logical intuition".' [Russell 1984, p. 101] This highly significant remark may well serve as one of the weaker links in Russell's logicist position whereby logic serves as the*rival* to intuition as a reliable foundation for mathematics. These observations carry even greater weight when one notes that that while Russell regards logic as *the* foundation for mathematics, he also chooses to characterize logic as "the study of the *forms* of complexes" [Russell 1992, p. 55, my emphasis]. In particular, based on Russell's latter characterization of logic, if intuition is indeed at the heart of our acquaintance with logic forms, it must also be foundational to logic.

But what might Poincaré contribute to our understanding of logical intuition? By way of a response, it of interest to observe that among his many representations of intuition is its status as that "faculty which makes us to see the end from afar" [Poincaré 1946a, p.218]. It is likely that what is required here, at least in more complex proofs, is Poincaré's idea, as Mooij expresses it, that "la mathématique doit faire appel ... continuelement á l'intuition". [Mooij 1966] This is perhaps the closest that Poincaré ever gets in his philosophical writings to identifying a possible role for logical intuition. As we are beginning to see, on contemplating this role alone, we can achieve no more than a glimpse of the scope that intuition has to play in explaining our acquaintance with logic and our identification of logically valid reasoning.

4. Poincare, Classical Logic and Finitism

While Poincaré considers the assumption of the actual infinite as being at the route of the set-theoretic antinomies, it is also noteworthy that he staunchly supports the principles of classical logic and in particular, *PEM*. Within the realm of constructive mathematics, this principle is typically denied the status of *a universal principle* on the basis of the understanding that truth is reducible to provability and that therefore a necessary and sufficient condition for $A \vee \neg A$ to be true is the availability of an effective rule for deciding which of

$A(x)$ and $\neg A(x)$ is true for any admissible value of x. As such, it should not be used as a basis for determining which objects exist within pure mathematics.

By contrast, for Poincaré, the need to re-define set membership at each step of construction of any set for which there is no predicative law of correspondence amounts to an unlawful violation of*PEM*. [Poincaré 1946b, p. 484; 1963, p. 45] The infallibility of this law is therefore a critical deciding factor for Poincaré in determining what constitutes a proper definition and in turn, which infinite sets must exist. It is as though he could not perceive of things being otherwise. Any attempt to conclude that his critique of logicism and formalism throughout his philosophical writings involves a disaffection towards the principles of logic ought to be carefully weighed up in the light of these observations, and indeed, his conclusion,

> "...What is, then, the condition for the rules of logic to be valid? It is this that the classification which is adopted be immutable." [Poincaré 1963, p. 45]

Poincaré's above position may therefore be regarded as semi-constructivist. In the first instance, infinite sets must be seen to be sets *constructible by an effective rule*. Since for Poincaré, this in turn requires the existence of a predicative law of correspondence, then *ipso facto*, only sets satisfying the requirements of classical logic exist. The constructive aspect of Poincaré's interpretation of set existence consists in a particular brand of finitism whereby sets must be surveyable entities in the sense we saw earlier. Here, surveyability has a special sense according to which, were it possible to observe the construction of an infinite set indefinitely, no previous members would require to be removed or to take up new positions at any step. This brand of finitism has important consequences for the status of the reals. In particular, Poincaré is persuaded to conclude that Cantor's refutation of the denumerability of the reals is not, as supposed traditionally, primarily a proof concerning the existence of a set of higher cardinality than $\boldsymbol{N}$; rather, it is in essence a proof of the absence of a predicative law of correspondence for constructing the reals, to the end that their very existence as an infinite set is refuted. [Poincaré 1963, p.62]

Poincaré allows his adherence to the classical principles of logic to take precedence in deciding which infinite sets exists, to the end that the reals *when construed as a set*,rather than strictly within the context of the arithmetization of the continuum, [Folina 1992] are fictitious.

He makes no mention of intuition, however, in defining the boundaries for proper sets. This leaves him in a rather peculiar position in relation to his logicist opponents. On the one hand, he uses classical logical as a basis for deciding which sets are admissible sets in pure mathematics. On the other hand, in denying logical reductionism, he stresses that logic perceived independently of the objects of pure mathematics is not logic proper. As far as Poincaré is concerned, to apply the rules of logic without specifying a domain of objects of pure mathematics would be to engage in a meaningless activity which is epistemologically impotent.

If, as Poincaré is suggesting here, classical logic and pure mathematics are inextricably linked, they ought to have a common foundation and this can be none other than the intuition of pure number. This belief in the commonality between the foundations of logic and pure mathematics (including the principle of mathematical induction) is particularly evident from his recognition of the intuition of pure number as that special variety of intuition which "directs" the "analysts" and his claims that,

> "It is by this procedure [mathematical induction] that the analysts have made science progress, and if we examine the detail itself of their demonstrations, we shall find it there at each instant beside the classic syllogism of Aristotle. ... It is the intuition of pure number, that of pure logical forms which directs those we have called *analysts*." [Poincaré 1946a, pp. 220 – 221].

By contrast, from Poincaré's perspective, logic, as presented by the logicists - "analytic logic"[3] – is "sterile" [Poincaré 1946b, p.483]. It is in a similar spirit, I believe, that Poincaré protests against logicism in 1905 by claiming that "in reducing mathematical thought to an empty form ... one mutilates it."[4]

From a Russellian standpoint, logic regarded as independent of its domain of application is not, as Poincaré would have it, empty. Rather, the key to the

---

[3]This denotation is likely to be intended by Poincaré in a derogatory sense and to have emerged from the dogma of logicism that pure mathematical statements are analytic. This dogma renders the intuition of pure number redundant in mathematical epistemology and allows the truth of the principles of logic to hold independently of an epistemological warrant in terms of their domain of application.

[4] This translation of remarks by Poincaré published in *1905, "Les mathématiques et logique", Revue de métaphysique et de morale 13, pp. 815-35* is taken from Goldfarb, W. (1988). Poincaré against the logicists. (In W. Aspray & P. Kitcher (Eds.), *History and Philosophy of Modern Mathematics*.(pp. 61 - 81) Minneapolis: University of Minnesota Press.) (p. 63).

content proper of logic can be found within the logical inferences themselves. The premises of such inferences carry all the required truth conditions for the conclusions to be true.[5] Thus, in the derivation of new propositions, logical inferences are truth preserving structures and as such, meaningful contentual structures. Moreover, the existence of inferences which give rise to distinguishable propositions affirms their value and is thought to undermine Poincaré's underlying theory [Poincaré 1946b, pp. 483-484] that logic, once applied to non-finitary predicates is sterile. Thus, Russell would have us believe that logic has a self-contained capacity for truth preservation and the generation of meaningful statements. If, as Poincaré seems to imply by his term *intuition of pure logical forms*, the proper source for these properties is intuition, then he offers a lack of compelling evidence to drive this point home. The need for a more careful consideration of the intuitive origin of logical forms is implicit not only as we have seen from Russell's writings but also from their traditional interpretation by John Cook Wilson (1849-1915), who was an ardent defender of Aristotelian logic, and by proponents of modern logic, as superior to the kinds of form derived purely from our natural language.

Folina [Folina 1996, p. 421] endeavours to rise to the challenge by noting in the first instance that Poincaré's intuition of pure number and intuition of pure logical forms are one and the same thing and may more properly be referred to as the "intuition of indefinite iteration." In so doing, she makes use of the logical principle *Modus Ponens*, which for logical formulae, *A* and *B*, can be formally expressed as a rule of inference as follows:

$$\frac{A \qquad A \rightarrow \mathrm{B}}{B} \quad . \tag{8}$$

However, Folina's approach is ultimately problematic in finding an intuitive basis for logic in so far as she assumes that we need *Modus Ponens*to convince ourselves that we are able to conceive of *PMI* as an indefinite iteration of syllogisms, and thus to affirm its intuitive status. [pp. 424 – 425] Later, [p. 426] Folina seeks to present the "intuition of pure iteration" as a *foundation* for logic by claiming that conceiving the indefinite iteration of an act is

[5] This containment relationship consists in the conclusions being true if and only if the premises in the corresponding inference are true.

assumed not only in the definition of "well-formed formula" but at a "deeper level" with the "bare ability to think logically or systematically". In particular, she claims that,

> "At a less conscious level [than understanding *why* a rule of inference is correct or proper], grasping a rule of logic minimally involves an implicit recognition of a pattern, a pattern of argument (or inference) that gets repeated in various contexts. Recognition involves some awareness of repetition; here it is repetition of the pattern once seen as possible, as justified. In either case, intuition is involved."

Thus, through taking logical reasoning to involve systematic thinking, and in turn, regarding systematic thinking as dependent on the ability to "recognize patterns, including some patterns of thinking". Folina goes on to conclude that "it is possible to argue here that this intuition must play a role in a general understanding of any rule of logic." The application of indefinite repetition required here is in essence the ability to conceive of a given rule of logic as having a warrant to be applied ad infinitum. Interestingly so, Folina uses *Modus Ponens* here as an illustration, thus conceding that we *need* the intuition of indefinite iteration to apply*Modus Ponens*, no less so than any other "rule of logic." However, her conclusion shortly thereafter that her above treatment of systematic thinking "is the ground for Poincaré's arguments for the epistemological priority of arithmetic over logic" begs the question when one recalls that she has also used logic earlier (in the form of *Modus Ponens*) as a foundational basis for *PMI* – a principle which is as Poincaré, himself argues, essentially *arithmetic*.

Poincaré's conviction regarding the intuitive status of *PMI* has no connection with logic in the above sense, however. Indeed, it is clear that in describing mathematical induction as "at once necessary to the mathematician and irreducible to logic" [Poincaré 1946b, p. 452] that he regards this principle as apodictic and foundationally independent of logic.

Thus, Folina's interpretation of Poincaré's intentions in using the term 'intuition of pure logical forms" lacks credibility in the light of Poincaré's own foundational treatment of *PMI*. Moreover, while it may be the case that Poincaré was sympathetic towards the notion of an intuition of indefinite iteration, it is evident from Parsons's work [Parsons 1980; 1983; 1986; 1993; 1998] that Poincaré's intuition of pure number enables him to attribute intuitive meaning and hence *immediate* "certitude" to a more complex level of mathematical thinking than indefinite iteration and in particular, can provide a foundational basis for arithmetic. [MacDougall 2010]

On the other hand, Folina is correct in assuming that Poincaré is also intent on refuting logical reductionism by appeal to intuitionas an indispensable foundation for the number words assumed in the formulation of well-formed formulae.(Folina 1996) The need for number words in the language of logic should not be confused, however, with our recognition of particular logical principles which are not intrinsically about number as infallible.

Grasping a logical principle as necessary is quite an independent exercise from that of applying the principle in an indefinite sequence of iterations. While Folina articulates this point, through subsuming Poincaré's intuition of pure logical forms within an intuition of definite iteration she fails to tackle the epistemological issues concerning Poincaré's affinity to particular principles of classical logic.If one were to attempt to argue that forPoincaré, indefinite iteration serves as the backbone to our recognition of anylogical principle as necessary, one ought to first consider (the Principle of) *Non-Contradiction* (*PNC*), which, for logical formulae A, may be formally expressed as,

$$\neg(A \wedge \neg A) \qquad (9)$$

In particular, one would be hard-pressed to explain his interpretation of this principle as "the skeleton, so to speak, of our intelligence" [Poincaré 1946b,p.430] – a description which is strikingly reminiscent of those of Aristotle, including "a principle which every one must have who knows anything about being", "the most certain of all" principles, [Aristotle 1984b, Metaph IV 3 1005^b15], "the most indisputable of all principles". [Metaph IV 4 1006^a5] and "the most indisputable of all beliefs" [Metaph IV 4 1011^b15].

As we shall see later, Poincaré's above treatment of *PNC* points to the unresolved need to acknowledge the place of logical intuition in equipping us with the essential principles of human reasoning.

5. Poincare's Notion of Pure Logical Form

But in what sense does Poincaré use the denotation 'pure logical forms' when referring to the intuition of pure number? I suggest that while Poincaré has not paid due attention to this question, what he intends here is akin to the notion I associated earlier with Russell's writings. As we saw earlier,

Poincaré's close affinity to Aristotelian syllogistic logic rather than the newer logics is evident from his 1905 comments. Thus, it is likely that by "pure logical forms" he specifically intends the structure of the sentences contained within valid arguments similar to Aristotle's first principles and syllogisms. Here, syllogisms constitute a special type of inferential structure involving two premises and a conclusion, as is illustrated in (7), above. Only these combinations of propositions which lead through deduction to *new* knowledge, and that of necessity, are viewed by Aristotle as permissible. [Smith 2007] Logical forms in the above sense are more precisely recognized via Aristotle's first principles and syllogisms as pure through his replacement of the subject and predicate terms by *schematic variables*, thus leading to the formation of *schematic forms*.

These forms are pure in a sense akin to that meant by Smith with reference to what is implicit from the writings of Poincaré's contemporary, Husserl in his "Prolegomena" [Husserl 1979]. Within this context, Smith suggests 'that "pure" logical form is form of judgement of expression abstracted from concrete acts of judging or speaking' [Smith 2003, p. 148]

In referring to Aristotle's work, the above interpretation of pure logical form is elaborated on by Menzel as follows:

> 'the notion of logical form is always relative to a theory T of logical formthat is directed toward some broad class K of sentences of natural language. Call K the *target class* of the theory. The logical forms of the members of K are represented by formulas in a "logically pure" canonical language, i.e., a language that contains (perhaps in addition to punctuation) only variables and logical constants - expressions with fixed and, according to the theory, distinctly logical meanings. For Aristotle, this is the language of schematic forms whose logical constants are "Every", "Some", "No", "is a", and "is not a"; and for Russell and his followers it is generally (some variant of) the language of Principia Mathematica [PM] with its now familiar quantifiers and propositional connectives.' [Menzel 1998]

This idea is also consistent with Kant's usage of the notion *logical form* according to which it is evident that he perceives purity as an essential quality of a logical form, as portrayed in the following remarks:

> "General logic. . . abstracts from all content of knowledge, that is, from all relationof knowledge to the object, and considers only the logical form in the relationof any knowledge to other knowledge; that is, it treats of the form of thought ingeneral." [Kant 1929, B80]

With reference to Kant's notion of abstraction in particular, MacFarlane captures more precisely what might be understood as occurring in the generation of logical forms by referring to the idea of "complete abstraction from semantical content" [MacFarlane 2002]. In turn, the *formality* of logic, as Kant intends it, consists in a) its separateness from "the semantic contents" – or as MacBeth would have it, the *empirical* [MacBeth 1994]contents of concepts and judgments and b) its focus on logical forms.

MacFarlane illustrates this quality as follows:

> "logic abstracts entirely from the fact that theconcept *cat* applies to cats and the concept *black* to black things, andconsiders only the way in which the two concepts are combined in thethought: the judgment's *form* (particular, affirmative, categorical, assertoric)". [MacFarlane 2002]

It is likely that Poincaré endorses Kant's interpretation of logical forms and his view that Aristotle presents logic in its entirety. However, the well-known limitations of Aristotelian logic in capturing all logical forms, particularly those with relational clauses, [Pietroski 2009] explains the added value gained through Russell's richer logical language and indeed at an earlier stage, that of Frege, in his *Begriffsschrift* of 1879, for deriving a logical apparatus for a formal system. Poincaré's "superficial acquaintance" [Jourdain 1912, p. 482] with more recent advances in logic within his own era and his associated concerns about the validity of a system with such freedom of expression may be preventing him from fully grasping this truth. [Couturat, 1912] Aristotle was able to demonstrate the invalidity of a proposed syllogism by means of a counterexample, thus demonstrating that the premises did not *syllogize*or equivalently, failed to produce a deduction. Where appropriate, he also had a method in place to confirm the validity of each of his syllogistic patterns either by verifying that the conclusion could not be falsified on assuming the premises (that is, by confirming the necessity of the conclusion) or directly through special rules, known as *conversions*. In the latter case, there are three possible conversions, each requiring the predicate and subject in the antecedent proposition to be reversed. [Smith 2007]

Aristotle's validity proofs maysuffice forPoincaré in relation to his dictum that to exist is "to be free from contradiction" [Poincaré 1946b, p. 474]. By contrast, the scope of the richer language of his contemporaries may prove rather formidable if this dictum is to constitute a diagnostic tool for logical validity. This interpretationof Poincaré is indeed supported by his untenable requirement for a formal system that,

> "it is … necessary to consider all the propositions deducible form these postulates considered as premises, and to show that, among these propositions, no two are contradictory."

and his failed attempt [Couturat 1912, p. 515] at appealing to *PMI* as a vehicle for achieving this end. [Poincaré 1946b, p. 454]

Thus, while the interpretation of "pure" as used by Poincaré is not necessarily in conflict with his recognition of objects of intuition as being closely tied to logic, it is implicit from his critique of the "new logics", including Russell's PM system that he lacks confidence that such systems can be justified on intuitive grounds to the same extent as Aristotelian logic. This need not excuse Couturat, however, from his recognitionof Poincaré's appeal to intuition in the derivation of the new logics as reducible to reliance on "simple common sense". [Couturat 1912]

With regard to the status of the intuition of pure number as an intuition of pure logical forms, Poincaré would have us believe that in pure mathematics the intuition of pure number enables us a) to construct surveyable objects in the sense referred to earlier and b) to present these objects to our understanding using logical forms. As such, the intuition of pure number enables us to decide what counts as an object of pure mathematics. Furthermore, to apply logical forms to unsurveyable objects is to disconnect these forms from their intuitive foundations in the absence of a sound epistemological warrant. Thus, Poincaré's usage of the term "pure logical forms" is more rightly a reflection of how logical forms are viewed by his contemporaries and in tracing their origin to the intuition of pure number, he is endeavouring to show that the objects of the intuition of pure number – the real objects of mathematics – and logical forms are tied together at their foundations. This position is reflected in his observations: "what it [mathematics] has gained in rigor, it has lost in objectivity. It is by separating itself from reality that it has acquired this perfect purity." [Poincaré 1946b, p. 435]

Nevertheless, there is an absence of evidence in Poincaré's work to convince us that pure logical forms should have their foundation in the intuition of pure number. Indeed, since Poincaré is notable for his recognition of varieties of intuition according to the discipline of interest, [MacDougall 2010] any attempt to strengthen his anti-logicist stance ought to involve consideration of the nature of *logical* intuition.

The idea of logical forms as being the basis for creating "secure" patterns of thought intended to guard against "falsity" is thought to explain their importance. [Pietroski 2009] It is this idea which might explain Poincaré'suse of the expression "[t]he rules of perfect logic" [Poincaré 1946b, p. 451] when referring to logic as a part of mathematics. This interpretation is consistent with Aristotle's usage of the term 'perfect' in relation to deductions when making the following distinction:

> "I call perfect a deduction which needs nothing other than what has been stated to make the necessity evident; a deduction is imperfect if it needs either one or more things, which are indeed the necessary consequences of the terms set down, but have not been assumed in the propositions." [Aristotle 1984a, Prior Analy I 1 24^{b}]

The above characteristics of logical deductions are also in keeping with the idea of appealing to our immediate inner sense and therefore in turn with the role of intuition as we understand it in the everyday sense. [MacDougall 2010] The idea that intuition – in particular logical intuition – is at work here would also help to give some justification for Poincaré's notion of logic as consisting of principles which are "common to all men, and that none can deny without being mad". [Poincaré 1946b, p. 383] The question of whether within Russell's new system of logic it is possible to differentiate between principles of logic which do and do not have a foundation in logical intuition is beyond the scope of this chapter. It is clear, however, from his own comments that Poincaré holds a particular affinity for *PNC* and *PEM* and it is difficult to see why this should be justified in terms of an intuition of *pure number*. Further, there is little credibility in the supposition that Poincaré was opposed to logic per se. Indeed, he refers to mathematics as "a science appealing only to the fundamental principles of logic", of which he reveals *PNC* to be an instance. [Poincaré 1946b, p. 430]

6. Embedding Logical Intuition within a More General Taxonomy for Intuition

To determine a logical form by means of logical intuition is to perceive corresponding propositions as concrete instances (or tokens) of that logical form. In this capacity, propositions are entirely similar to Parsons's quasi-concrete objects, including strings of strokes and geometric figures in the case of number theory and geometry, respectively. This is evident from considering the following definition referring to the latter types of quasi-concrete object:

> "To determine a quasi-concrete object by means of [intuition] is to perceive concrete instances(or, tokens) of that object as its types." [MacDougall 2010, p. 145]

Here, however, I extend the application of the above definition by observing that logical forms are themselves objects of logical intuition – a special case of Parsons's quasi-concrete objects. Logical forms carry a further defining characteristic of other types of quasi-concrete objects in that while they are objects in their own right they are also intrinsically related to concrete objects. These relationships are realized through the instantiation of the variables in purely formal expressions by expressions referring to concrete objects.

The case has already been defended elsewhere for the use of Parsons's structuralism based on his theory of quasi-concrete objects, to warrant a more harmonious approach to intuition for explaining the foundations of arithmetic and geometry and beyond than that previously posited by Poincaré. [MacDougall 2010] Taking the relations expressed by logical forms to be particular structures, the above proposal is now extended to include the foundations of logic. However, this is with the understanding that logic is not a branch ofmathematics and that therefore logical intuition is not a special type of mathematical intuition. Further, the discovery of the valid principles of logic is grounded on our sensory experience of physical objects, not, as Poincaré would have it, on our study of mathematical objects.

More generally, taking logical intuition to be a faculty which enables us to accept particular logical relations between concrete objects, just as in the case of geometric intuition and the intuition of pure number, [MacDougall 2010] the physical world and the objects it contains still have a prominent epistemological role to play. In particular, they are required to exercise this faculty and indeed, to bring its existence to our attention, but not to justify the existence of the logical forms which we use to attribute structure to the relation between physical objects.

From this perspective, the principles of classical logic may be regarded as a framework for interpretation of the physical world. As such, foundationally speaking, they are neither a) derived through abstraction from pre-existing relations between physical objects nor b) derived from the declarative sentences of natural language. While as Frege and more explicitly, Quine, would have it, the language of logic may be thought of as a "regimented" version of natural language characterized by clarity, [Shapiro 1998] this does not require that natural language is anterior to logic.

Logical intuition is necessary both for the identification of valid axioms and rules of deduction for a given deductive apparatus and for the perception of these axioms and rules as valid. It is responsible for enabling us to recognize the logical form for a simple proposition and a given argument pattern representing a rule of inference as logically valid. In allowing us to see the end from afar, logical intuition, should also enable us to embrace a given proof structure on intuitive grounds. While, as we have already seen, Poincaré, points us to this requirement for intuition more generally, Menzel captures what is required here more accurately by means of the following remarks:

> "The argument's intuitive validity is then explained in virtue of the fact that the logical forms of the premises formally entail the logical form of the conclusion. The primary goal of a theory of logical form is to explain as broad a range of such intuitive logical phenomena as possible in terms of the logical forms that it assigns to sentences of natural language." [Menzel 1998]

From a structuralist perspective, analogies between logic and geometry have previously been proposed, although not specifically within the context of discussions on intuition. In particular, Wittgenstein describes logic and geometry as each "[describing] *form* rather than content" [Summerfield 1990] Further, Wittgenstein's recognition of the necessity of logical forms is encapsulated by his remarks that,

> "2.18 What any picture, of whatever form, must have in common with reality, in order to be able to depict it – correctly or incorrectly – in any way at all, is logical form, i.e. the form of reality."

and

> "3.031 It used to be said that God would create anything except what would be contrary to the laws of logic. The truth is that we could not say what an 'illogical' world would look like." [Wittgenstein 1974]

In so far as Frege viewed the logical form of a proposition as the *manifestation* of the structure of a sentence, the above application of Parsons's structuralism to logic also forms a meeting ground with Frege, without the need to embrace his logicism [Pietroski 2009]

Contrary to what Couturat would have us believe, logical intuition in the above sense is also more than "simple common sense". Additionally, by uniting logic and intuition, it offers a precise alternative to Couturat's view that those who appeal to logic "are less exposed to error" than those who appeal to intuition [Couturat 1912, p. 491], thus strengthening Poincaré's anti-logicist cause. This is particularly important when one considers Couturat's conclusion that,

> "the vague conception of *intuition* is out of place as a weapon against the logicians, especially when the intuition spoken of is not specified [by Poincaré]." [Couturat 1912, p. 497]

On the other hand, much can be gained from rendering the language of logic richer through accommodating a wider range of forms of reasoning than is permissible on the basis of Aristotelian logic alone. Thus, contrary to what Poincaré would have us believe [Couturat 1912], the contributions to the formalization of logic made by his contemporaries Peano and Russell need not be viewed as a threat to the foundational role of intuition in the derivation of logic. Rather, many of these developments ought to be regarded as a means of arriving more precisely at what is conveyed through logical intuition, thus *facilitating* invention in mathematics.

Assuming that logical intuition is a faculty with which we have been endowed to enable us to engage with the physical world as an objective world and that the domain of mathematics is fundamentally a construction of the human mind, then it may be argued that logical intuition, unlike the intuition of pure number, is fallible in its capacity to contribute truth and meaning to pure mathematics. From this standpoint, it can be argued that in order to guarantee genuine productivity *in pure mathematics*, *a priori* knowledge acquired through logical intuition must be counterpoised by that acquired through the intuition of pure number. This position is close to that of Poincaré's in his recognition of intuition as "counterpoise" to logic, [Poincaré 1946a, p. 217]. If one probes a little deeper it would also appear to be consistent with Aristotle's perspective on the foundations of mathematics. In his *Posterior Analytics,Aristotle* offers a brief account of the origin of the first principles of science. In so doing, he recognizes them as those principles of which we have knowledge without demonstration, with which we are "more familiar" [Aristotle 1984b, Post. An. I.2.] and from which we must be able to derive all other propositions. Here the need for the cognitive faculty *Nous*is acknowledged.

For Aristotle, *Nous* explains why elementary principles seem familiar to us prior to any scientific demonstration. Thus, in particular, he views Nous as a "capacity or power (*dunamis*)" [Smith 2007] to recognize certain logical principles as necessary, just asour senses to recognize colours without the need of *a priori* demonstration.

It is clear that Aristotle recognizes the necessity of sense objects for identifying this faculty. If, as has been suggested elsewhere, Nous is (intellectual) intuition, [Smith 2007] then it can be argued that the realization of logical principles through the interaction of intuition with physical objects serves as a basis for explaining Poincaré's affinity to classical logic. Indeed, his portrayal of intuition as a faculty, in the sense of a mechanism or power, responsible for feelings of certitude concerning non-verifiable principles [MacDougall 2010] is very reminiscent of Aristotle's representation of Nous as described above. Thus, in particular, Poincaré's recognition of *PEM* as a necessary principle of mathematics and *Non-contradiction* as "the skeleton, so to speak, of our intelligence" [Poincaré 1946b, p. 430] are likely to have their foundation in logical intuition as a faculty for identifying *a priori* principles which apply to the physical world.

Clearly, however, not all intuitive knowledge is mathematical knowledge, and one is at liberty therefore to ask whether Aristotelian and later, Russellian logic ought to be treated as such by Poincaré. What is more likely is that logical intuition provides us with the essential foundational platform from which to commence our search for the true nature of mathematical proof and certitude. From this perspective, it is important to acknowledge the fundamental distinction between truth in an objective physical world and truth in a mind-dependent mathematical world. Understood in this way, logical intuition may be recognized as a special variety of intuition satisfying the interpretation of intuition proposed by MacDougall [MacDougall 2010], whereby the structures and relations presented by sensibility are in this case captured by individual logical forms and the objects of reason which satisfy the relations are in the first instance, physical objects presented to our minds through the lens of sensibility. Such insights serve as a comprehensive anti-logicist justification forRussell's hitherto neglected discoveries that logical forms areentities with which we have acquaintance and which constitute objects and structures in their own right.

REFERENCES

Aristotle (1984a). *The Complete Works of Aristotle (Volume 1): The Revised Oxford Translation.* (J. Barnes, ed., Princeton, New Jersey: Princeton University Press)

Aristotle (1984b). *The Complete Works of Aristotle (Volume 2): The Revised Oxford Translation.* (J. Barnes, ed., Princeton, New Jersey: Princeton University Press)

Bishop, E. (1967). *Foundations of constructive mathematics.* (New York: McGraw-Hill)

Chrystal, G. (1904). *Algebra: An Elementary Textbook.* (London: Adam and Charles Black)

Couturat, L. (1912). For Logicists: Reply to M. Poincaré (First published 1906, Philip E B Jourdain, Trans.). *The Monist*, *XXII*, 483 - 523

Feferman, S. (1964). Systems of predicative analysis. *Journal of Symbolic Logic*, *29*, 1 - 30

Folina, J. (1992). *Poincaré and the philosophy of mathematics.* (London: MacMillan)

Folina, J.(1996). Logic and intuition in Poincaré's philosophy of mathematics.(In J.L. Greffe, G. Heinzmann & K. Lorenz (Eds.),*Henri Poincaré: Science and philosophy.* (pp. 417 - 434)Berlin: Akademie Verlag.)

Frege, G. (1964). *The Basic Laws of Arithmetic: Exposition of the System (partial translation of Vol. 1 and Vol. 2 of the German texts of 1893 and 1903, resp.).* (Berkeley and Los Angeles, USA: University of California Press)

Gödel, K.(1944).Russell's Mathematical Logic.(In P.A. Schlipp (Ed.),*The Philosophy of Bertrand Russell, vol. V.*(pp. 123 - 153)London: Cambridge University Press.)

Gödel, K. (1962). *On Formally Undecidable Propositions of Principia Mathematica and Related Systems (B. Meltzer, Transl. with introduction by R.B. Braithwaite.).* (Edinburgh: Oliver and Boyd (First published 1931))

Goldfarb, W.(1988).Poincaré against the logicists.(In W. Aspray & P. Kitcher (Eds.),*History and Philosophy of Modern Mathematics.*(pp. 61 - 81)Minneapolis: University of Minnesota Press.)

Heinzmann, G.(2006).Henri Poincaré and His Thoughts on the Philosophy of Science (In É. Charpentier, É. Ghys & A. Lesne (Eds.),*The Scientific Legacy of Poincaré.*(pp. 373 - 391)Providence, Road Island, USA and

London, England: American Mathematical Society and London Mathematical Society.)

Holmes, M.R. (2006) Alternative Axiomatic Set Theories. *Stanford Encyclopedia of Philosophy*. Stanford, USA: Metaphysics Research Lab, Stanford University.

Holmes, M.R. (2007) Pocket Set Theory. Boise, Idaho, USA: Boise State University.

Husserl, E. (1979). *Logical Investigations (Vol. 1)*. (London: Routledge and K Paul (first published 1900))

Jones, O. (1995) Nous. In T. Honderich, *The Oxford Companion to Philosophy*. Oxford: Oxford University Press, p. 628.

Jourdain, P. (1912). For logicists: Introductory Note. *The Monist*, *XXII*, 481 - 483

Kant, I. (1929). *Immanuel Kant's Critique of Pure Reason* (London: MacMillan. (First published 1787))

Łukasiewicz, J. (1957). *Aritstotle's Syllogistic from the Standpoint of Modern Logic*. (Oxford: Clarendon Press)

MacBeth, D. (1994). The Coin of the Intentional Realm. *Journal for the Theory of Social Behaviour*, *24*, 143 - 166

MacDougall, M. (2010). Poincaréan intuition revisited: what can we learn from Kant and Parsons? *Studies in History and Philosophy of Science*, *41*, 138 - 147

MacFarlane, J. (2002). Frege, Kant and the Logic in Logicism. *The Philosophical Review*, *3*, 25 - 65

Mares, E.D. (2007). Russell's Logical Forms. *Soochow Journal of Philosophical Studies*, *16*, 215 - 256

Menzel, C.(1998).Logical form.(In E. Craig (Ed.),*Routledge Encyclopedia of Philosophy, vol.* 5, London and New York: Routledge.)

Mooij, J.J.A. (1966). *La Philosophie des Mathématiques de Henri Poincaré*. (Paris: Gauthier-Villars)

Parsons, C. (1980). Mathematical Intuition. *Proceedings of the Aristotelian Society*, *80*, 145 - 168

Parsons, C. (1983). *Mathematics in Philosophy: Selected Essays*. (Ithaca, NY: Cornell University Press)

Parsons, C.(1986).Intuitionin Constructive Mathematics.(In J. Butterfield (Ed.),*Language, Mind and Logic*.(pp. 211 - 229)Cambridge: Cambridge University Press.)

Parsons, C.(1991).The impredicativity of induction.(In M. Detlefsen (Ed.),*Proof, Logic and Formalization*.(pp. 139 - 161)London: Routledge.)

Parsons, C. (1993). On Some Difficulties Concerning Intuition and Intuitive Knowledge. *Mind*, *102*, 233 - 246

Parsons, C.(1998).Finitism and Intuitive Knowledge.(In M. Schirn (Ed.),*The Philosophy of Mathematics today*.(pp. 249 - 270)Oxford: Clarendon Press.)

Pietroski, P. (2009) Logical Form. *Stanford Encyclopedia of Philosophy*. Stanford, USA: Stanford University

Poincaré, H. (1909b). Réflexions sur les Deux Notes Précédentes *Acta Mathematica*, *32*, 195 - 200

Poincaré, H.(1946a).The Value of Science.(*The foundations of science*.(pp. 201 - 358)Lancaster, PA: The Science Press.)

Poincaré, H.(1946b).Science and Method.(*The foundations of science*.(pp. 359 - 547)Lancaster, PA: The Science Press.)

Poincaré, H.(1963).The logic of infinity.(*Mathematics and science: Last essays*.(pp. 45 - 64)New York, USA: Dover Publications, Inc.)

Russell, B. (1911). The Philosophical Importance of Mathematical Logic (P. Jourdain Transl.). *The Monist*, *23*, 481 - 493

Russell, B.(1967).Mathematical Logic as Based on the Theory of Types (originally published 1908).(In J. van Heijennort (Ed.),*From Frege to Gödel: A Source Book in Mathematical Logic*.(pp. 150 - 182)Cambridge, England: Harvard University Press.)

Russell, B.(1984).Theory of Knowledge (In E.R. Eames & K. Blackwell (Eds.),*The Collected Papers of Bertrand Russell, vol. 7*.(pp. 1 - 178)London: Routledge (First published 1913).)

Russell, B.(1984a).The Philosophical Importance of Mathematical Logic (originally published in 1911).(*The Collected Papers of Bertrand Russell, vol. 7*.(pp. 1 - 178)London: Routledge.)

Russell, B.(1992).What is Logic?(In J.G. Slater & B. Frohmann (Eds.),*The Collected Papers of Bertrand Russell, vol. 6*.(pp. 54 - 56)London: Routledge (First published 1912).)

Russell, B. (1993). *Introduction to Mathematical Philosophy*. (London: Routledge (First published 1919))

Shapiro, S.(1998).Logical Consequence: Models and Modality.(In M. Schirn (Ed.),*The Philosophy of Mathematics Today*.(pp. 131 - 156)Oxford: Oxford University Press.)

Smith, D.W. (2003). "Pure" logic, ontology, and phenomenology. *Revue internationale de philosophie*, *2*, 21 - 44

Smith, R. (2007) Aristotle's Logic. *Stanford Encyclopedia of Philosophy*. Stanford, USA: Stanford University

Summerfield, D. (1990). Wittgenstein on logical form and Kantian geometry. *Dialogue*, *XXIX*, 531 - 550

Tiles, M. (1989). *The Philosophy of Set Theory: an Historical Introduction to Cantor's Paradise*. (Oxford: Basil Blackwell)

von Wright, G. & Anscombe, G. eds. (1961). *Notebooks*. New York: Harper and Row.

Whitehead, A.N. & Russell, B. (1925). *Principia Mathematica (Vol. 1)*. (Cambridge: Cambridge University Press (First published 1925))

Wittgenstein, L. (1974). *Tractatus Logico-Philosophicus (D.F. Pears & B.F. McGuinness, Transl. with introduction by B. Russell)*. (London: Routledge (First published 1921))

Zermelo, E.(1908).A new proof of the possibility of a well-ordering. (In J. van Heijenoort (Ed.), From *Frege to Gödel: A Source Book in Mathematical Logic*.(pp. 185 - 215)Cambridge, Massachusetts: Harvard University Press.)

In: Philosophy of Science
Editors: S. Pintuck and C. Reynolds
ISBN: 978-1-62100-276-5

Chapter 3

DIFFERENT DOMAINS, THE SAME PROBLEMS

***Martín Narvaja*[1]*, Mariana Córdoba*[2] *and Olimpia Lombardi*[3,*]**
[1,2]CONICET – FFyL, Universidad de Buenos Aires, Argentina
[3]CONICET – FCEyN and FFyL, Universidad de Buenos Aires, Argentina

ABSTRACT

In this chapter, we argue that both the problem of quantum indistinguishability in the philosophy of physics and the problem of identity in structures in the philosophy of mathematics are identical in a certain relevant sense; i.e., that they are two instances of the same general problem. This favors the idea that certain ontological questions are common to different domains, in spite of the specificity of each domain; hence, the solutions proposed in one of the domains may be extrapolated and adapted to the others.

1. INTRODUCTION

The search for unity behind multiplicity was a central theme in western philosophy since its birth in Ancient Greece. This line of thought was inherited

* Corresponding author: olimpiafilo@arnet.com.ar

by Modern science, through the idea that there are some common aspects underlying the multiplicity of phenomena. The philosophy of science of the twentieth century adapted the same idea to the epistemological realm: in spite of the specific features of physics, biology, chemistry, and other scientific disciplines, certain problems related with, say, realism, explanation, reduction or induction can be found in all of those sciences.

The purpose of this chapter is to show that the philosophy of science can provide an analogous insight regarding ontological issues. We will claim that certain apparently unrelated difficulties, that come from different disciplines, are actually identical. In particular, we will argue that the problem of indistinguishability of quantum particles –that belongs to the philosophy of physics– and the problem of identity of places in the structuralist account of mathematical reality –an issue of the philosophy of mathematics– are two examples of a single ontological problem. Moreover, this way of conceiving seemingly disconnected problems favors the idea that the solution proposed in one of the domains may be extrapolated and adapted to the others.

Usually, a number of different issues belonging to the specific fields of mathematics and physics are considered in the discussions about the problems mentioned above. Nevertheless, the Principle of the Identity of Indiscernibles, or Leibniz's Law, is involved in all of them. The principle states that if an object or individual has a property as long as another object or individual has, and *vice versa*, and this applies to all of their properties, then both objects are the same object. That is to say, the allegedly two objects are just one and the same. Formally, being x and y universally quantified variables of individuals,

$$\forall P\left(Px \leftrightarrow Py\right) \rightarrow x = y \qquad (1)$$

In this way, Leibniz's Law establishes a sufficient condition for the identity of objects. It shall be noted that when the range of P increases, the domain of the considered properties also increases. And this implies that the condition in the left part of the law increases its strength and, then, is harder to be satisfied; consequently, the principle becomes weaker. There are different versions of the principle depending on which properties are considered (French & Krause 2006). According to the weaker version, it is not possible for two objects to be indiscernible regarding all of their properties and relations. A second version excludes the space-time properties from the former set. The stronger version also excludes relational properties. According to the last version, it is impossible for two objects to be indiscernible in all of their monadic properties. In the next two sections we will show that the requirements of Leibniz's Law are neither satisfied in quantum mechanics nor in mathematical structuralism.

The chapter is divided into three sections. First, we will discuss the problem of quantum indistinguishability: How the indistinguishability of quantum particles must be conceived? Is it a problem open to further physical enquiries or is it only a matter of philosophical interpretation? (see, *e.g.*, Castellani 1998, French & Krause 2006). Second, we will turn our attention to the field of the philosophy of mathematics, where we will summarize the structural proposal of Shapiro (1997) and the problem of the identity of places in structures noticed by Keränen (2001). Finally, we will argue that, in spite of the different nature of the domains to which these two problems belong, they are essentially the same problem. As a consequence, we will suggest a single strategy for facing the solution in both cases.

2. The Problem of Indistinguishability on Quantum Mechanics

The problem of quantum indistinguishability finds its roots in quantum statistics and is as old as quantum mechanics itself. It can be stated in simple terms: quantum systems seem to violate Leibniz's Principle of Identity of Indiscernibles, that is, they may be only numerically different (French & Krause 2006). Of course, this requires some further explanations; the present section is devoted to supply them.

When two or more particles share their intrinsic or characteristic properties, these particles are called "identical" (van Fraassen 1998). Two silver atoms or two electrons are identical. Indeed, all of the atoms and systems of the same type are identical in this sense, that is, regarding their intrinsic properties. Nevertheless, in the classical context, they are distinguishable by their dynamical or extrinsic properties (*i.e.*, position and momentum). In other words, they can always be distinguished by their states and identified by its space-time properties (Jauch 1968, p. 276). But, as we shall see, Leibniz's Law faces a deep challenge in the context of quantum mechanics.

According to the Maxwell-Boltzmann statistics, given N classical particles to be distributed over M energy levels, there are $W_C = M^N$ possible distributions (*complexions*). In particular, if we consider two systems (say A and B) of the same nature ("identical particles") to be distributed over two energy levels (say E_1 and E_2), there are four possible combinations: in spite of the intrinsic indistinguishability of the two component systems, the principle of individuation, no matter which one, makes particle A in level E_1 and particle B in level E_2 a different combination than particle A in level E_2 and particle B in level E_1 (see Table 1).

Table 1. Distribution of two classical systems over two energy levels

	E_1	E_2
1	AB	
2	A	B
3	B	A
4		AB

When our statistics are computed over quantum systems, the result is different. The origin of quantum statistics can be traced back to the works of Planck on the black body radiation (Singh 2005). Planck assumed that the total energy of the body was quantized because absorbed/emitted by oscillators in the walls of the cavity, and that those oscillators could only exchange discrete quantities of energy given by $\varepsilon = h\nu$ (the minimal quantum of energy). The total energy of the system is, then, $U = N\varepsilon = Nh\nu$, being N the number of energy elements. Planck computed the number of the possible distributions of the N energy elements $\varepsilon = h\nu$ over M oscillators, and he found that, instead the value W_C of the classical statistic, the right value was W_Q:

$$W_Q = \frac{(M+N-1)!}{N!(M-1)!} \tag{2}$$

If we distribute two energy elements ($N = 2$) between two oscillators O_1 and O_2 ($M = 2$) according to Equation (2), we find that there are only three possible complexions (see Table 2).

Table 2. Distribution of two energy elements over two oscillators

	O_1	O_2
1	**	
2	*	*
3		**

In 1924, inspired by a work received from Satyendra Bose, Einstein developed a method to count possible states of aggregation for photons on the basis of the equation (2); this led him to the so-called Bose-Einstein statistics. This result was later extended far beyond photons, to all integer-spin particles. The fundamental consequence that Einstein extracted from Bose's work was that, when we talk about entities governed by quantum statistics, we talk about entities that are strictly indiscernible, that is, entities that can be only numerically different (Krause 2006). This means that if two quantum systems have to be distributed over two possible states, there are only three possible complexions; as in the case of Planck's energy elements (see again Table 2).

Two years later, in 1926, Fermi and Dirac deduced a second kind of quantum statistics, which applies to systems of half-integer spin, like electrons. These statistics obey Pauli's Exclusion Principle. In spite of the difference between the two quantum statistics, Bose-Einstein for bosons (integer-spin particles) and Fermi Dirac for fermions (half-integer-spin particles), the problem regarding indistinguishability is the same. In the following paragraphs we will focus on bosons; nevertheless, the same considerations can be extended to fermions (see Ladyman & Bigaj 2010).

We have said that, in classical statistical physics, when we have two systems A and B to be distributed over two states, E_1 and E_2, we have four possible combinations (Table 1). It is also assumed that each complexion has the same probability: $\Pr(1) = \Pr(2) = \Pr(3) = \Pr(4) = 1/4$. We have also said that the result is different in the quantum case: according to Bose-Einstein statistics, in an analogous situation (two systems to be distributed over two states) we have just three complexions. In fact, complexions 2 and 3 from

Table 1 are counted as only one in Table 2. However, in this case, we do not consider that the new "combined" complexion has a higher probability, precisely, $\Pr(2\text{-}3) = 1/4 + 1/4 = 1/2$; on the contrary, all complexions are considered again as equally probable: $\Pr(1) = \Pr(2) = \Pr(3) = 1/3$.

The usual way of reproducing these statistics in the quantum formalism proceeds by beginning with systems as if they were distinguishable, and by then erasing their individuality (on this point, we follow Ballentine 1998). Let us consider a system composed by two systems –particles– A and B, represented in a Hilbert space $\mathrm{H}_{\mathrm{AB}} = \mathrm{H}_{\mathrm{A}} \otimes \mathrm{H}_{\mathrm{B}}$, where H_{A} and H_{B} are the Hilbert spaces corresponding to the component systems. If A and B are in the pure states $|\Psi_1\rangle \in \mathrm{H}_{\mathrm{A}}$ and $|\Psi_2\rangle \in \mathrm{H}_{\mathrm{B}}$ respectively, then the state of the composite system is $|\Psi_{12}\rangle = |\Psi_1\rangle \otimes |\Psi_2\rangle \in \mathrm{H}_{\mathrm{AB}}$; if A and B are in the pure states $|\Psi_2\rangle \in \mathrm{H}_{\mathrm{A}}$ and $|\Psi_1\rangle \in \mathrm{H}_{\mathrm{B}}$ respectively, then the state of the composite system is $|\Psi_{21}\rangle = |\Psi_2\rangle \otimes |\Psi_1\rangle \in \mathrm{H}_{\mathrm{AB}}$. The results of quantum statistics imply that states $|\Psi_{12}\rangle$ and $|\Psi_{21}\rangle$ represent the same physical situation. Therefore, the permutation of the systems A and B does not lead to a new possibility. Strictly speaking, the vectors $|\Psi_{12}\rangle$ and $|\Psi_{21}\rangle$ must represent the same physical state.

This idea has a general expression in the Indistinguishability Postulate (Ballentine 1998, p. 473): If a system S is composed by N "identical" subsystems $S_1, S_2, ..., S_N$, then the dynamical states of S that differ only by the permutation of the identical subsystems cannot be distinguished. As a result of this requirement, the permutations of identical subsystems lead to symmetrical states, corresponding to the Bose-Einstein statistics, or to anti-symmetrical states, corresponding to the Fermi-Dirac statistics.

In this formal strategy, the "names" of the systems (and the assumption of their individuality) are introduced just for formal purposes; but later they are erased (with their individuality) by means of the Indistinguishability Postulate (Krause 2006). Here, Leibniz's Law is violated even in its weaker formulation. This means that quantum systems are indistinguishable not only regarding their intrinsic properties, but also with respect to their dynamical and space-time properties.

Two philosophical perspectives have been proposed to face the quantum indistinguishability problem: the "non-individual" view and the "inaccessible-states" view. According to the first one, quantum entities are not objects or individuals at all; therefore, Leibniz's Law not only fails for quantum entities, but it does not even apply to them (Post 1963). This perspective, also called "inherited conception" (French & Krause 2006), justifies the superfluous nature of names and labels for quantum systems by appealing to the idea that, in the quantum realm, the talk about individuals that could be named, labeled, or even identified is meaningless (Post 1963, Teller 1983, 1995, 1998). Labels cannot be attached to quantum systems because they do not possess "haecceity", that is, what makes an object to be different from all others in some way that transcends all properties (Teller 1998).

How should we understand quantum statistics according to the "non-individual" view? Hesse (1963) and Teller (1998) suggest an analogy: the distribution of quantum systems in states is very much like the distribution of money in bank accounts. If we ask how to distribute two pounds in two bank accounts, we have just three possibilities: one pound in each account, both in the first account or both in the second one. It would be meaningless (and awkward) to say that this pound goes here and this other pound goes there. Nevertheless, the analogy has its obvious limitations. Are we saying that quantum particles are as abstract as money? And if we accept this, which kind of abstract entities are they? Whichever answer we give to these questions, a problem remains: classical logic and set theory are built upon an ontology of individuals and obey Leibniz's Law (at least in its weakest formulation). For this reason, a family of logics has been proposed in order to treat this kind of non-individual entities (da Costa & Krause 1994, 1997, 1999, Krause 2010).

From the second, "inaccessible-states" perspective (French & Redhead 1988, Redhead & Teller 1992, van Fraassen 1998), the fact that labels are superfluous from the physical point of view does not force us to accept that they cannot be metaphysically attached to individual objects. In other words, although quantum entities cannot be individually identified, they could preserve their metaphysical individuality, individual essence or quiddity. The compatibility between this assumption and quantum statistics can be obtained by conceiving situations physically equivalent but metaphysically different. This forces the postulation of a strictly metaphysical domain only for philosophical purposes, and the rejection of certain assumptions in the probability theory (not all of the metaphysical complexions are equally probable). Perhaps, as a consequence of this physics-metaphysics duality, this position needs the assumption that the states resulting from the permutation of "identical" systems, although ontologically possible, do not occur physically: quantum statistics is recovered by regarding those states as possible but never actually realized.

From this point of view, the Indistinguishability Postulate is considered a physical principle that imposes restrictions over the states accessible to quantum systems: once a system "gets into a subspace", he can't get out (the permutations would be possible but never realized observables). This kind of symmetry (symmetry under permutations) is interpreted as an additional constant of motion of the composite system (French & Krause 2006, Redhead & Teller 1992). However, this perspective has received a number of criticisms (Huggett 1995). On the one hand, the only reason to introduce this new kind of constant of motion seems to be our difficulty to accept indistinguishability in our ontology. Moreover, the postulation of a physical hypothesis for merely philosophical reasons has a strong ad hoc flavor. On the other hand, no further physical enquiries have been, or could be, developed on the basis of this assumption.In summary, we seem to be trapped in a dilemma: if labels can be applied in the quantum world, then we have to accept that there are possibilities that never become actual; if, on the contrary, labels make no sense for indistinguishable particles, we are committed to develop a reasonable conception of entities without identity and a logic for them.

With respect to this dilemma, and inspired by structural realism (Worrall 1989), van Fraassen (2006) argues that the traditional perspectives on indistinguishability have missed the point. He claims that quantum mechanics only characterizes structures, which can be fulfilled with different kinds of items; however, which particular kind of items is selected for this task is completely irrelevant. According to van Fraassen, the only relevant content of quantum mechanics is that reality has some structural properties. It is interesting to notice that the logic that seems to be adequate to talk about such structures is that developed by Krause (2005) to treat quantum systems. In this logical framework, structures behave as quantum composite systems, and places in the structure play the role of "quantum particles".

3. The Problem of Places in Mathematical Structuralism

One of the main questions in the philosophy of mathematics is that referred to the nature of mathematical entities. According to Platonism, mathematical entities are objects that inhabit a non-empirical but real domain, and their properties and relations can be discovered by studying that domain. Conventionalists, on the contrary, conceive mathematics as the activity of developing formal systems, which may be useful in different contexts and for different purposes; the question about the nature of mathematical entities makes sense only in this conventional framework.

Mathematical structuralism intends to supply an alternative to both conventionalism and Frege's or other versions of Platonism. It is inspired in Dedekind's idea that mathematics characterizes its objects up to isomorphism, and it is clearly exposed in the classical and controversial paper by Benacerraf (1965), "What numbers could not be?" The main notion in the different versions of mathematical structuralism is, of course, that of structure, whose essential feature is its possibility of having different instances. In other words, two or more systems may have the same structure, as two buildings may have been built with the same plan (Shapiro 1997).

On this general basis, the variety of structuralist proposals differ from one another with respect to the status attributed to structures. There are three main perspectives on this point: the *in re* version of Resnik (1997), the modal version of Hellman (1996) and the *ante rem* version of Shapiro (1997). Both the *in re* and the modal version are, as Shapiro argues, structuralisms without structures, because –in different ways– they finally eliminate structures in favor of a more fundamental category (in general, the category of set). We will begin with presenting Shapiro's *ante rem* structuralism and some of its ontological implications, in order to understand the discussion triggered by this proposal.

The main feature that any structure must possess is that of multiple exemplification: it must be possible to assert that two systems have the same structure. On the other hand, according to structuralism, in any mathematical theory, numbers are nothing but places in a structure: the search for the identity of a number considered in isolation –*i.e.*, regardless of the other members of the structure– makes no sense. For instance, Shapiro agrees with Benacerraf in considering that arithmetic is a science about structures, and whatever a natural number is –its "essence"– is its relation to other natural numbers: "The number 2, for example, is no more and no less than the second position in the natural-number structure (...). The essence of 2 is to be the successor of the successor of 0, the predecessor of 3, and so on." (Shapiro 1997, p. 72).

With the explicit purpose of defending ontological realism, Shapiro claims that two ontological questions must be asked, one concerning structures and the other concerning the places in a structure. Regarding the ontological status of structures, he defends an *ante rem* realism: structures are freestanding, they are independent of the systems that instantiate them. In other words, structures have ontological priority over their places. And since structures themselves "are in the ontology" (Shapiro 1997, p. 92), Shapiro argues that *ante rem* structuralism requires a theory of structures, with a precisely defined identity relation between structures. With this purpose, Shapiro proposes an axiomatic development of Zermelo-Fraenkel set theory in a second-order language, where identity among structures is adopted as a primitive notion, being isomorphism a congruence among structures: in this theory, it is stipulated that two structures are identical if they are isomorphic.

However, the question that still needs to be answered is whether a place in a structure is an object or not. At this point, the positions of Shapiro and Benacerraf begin to diverge: whereas Benacerraf considers that numbers are not objects at all, they are not objects *in any sense*, Shapiro seems uncomfortable with Benacerraf's conclusion since it does not agree with mathematical realism. The challenge of Shapiro is to develop a philosophical framework where numbers –which are no more and no less than places in structures– are also objects and have –in some way– an independent existence.

With this purpose, Shapiro distinguishes between the *places-are-offices* perspective and the *places-are-objects* perspective. According to the former, the positions in a structure are discussed in the context of one or more systems that exemplify the structure. They constitute places in the sense that they must be fulfilled with certain objects given by a certain logically previous ontology. This means that the *places-are-offices* perspective presupposes a background ontology that supplies those objects that will fill the places of the structure. From this point of view, we are dealing with places in virtue of the objects that instantiate them. From the *places-are-objects* perspective, on the contrary, places in a structure are treated as genuine, *bona fide* objects. These objects constitute the referents of certain terms –having restricted the language to the theory of the structure, that is, to a language limited to the terms of the places and the relations of the structure under consideration. According to this perspective, places are considered in themselves, regardless of the exemplifications that they could have: 2 is the successor of the successor of 0, independently of which objects instantiate that structure.

In summary, the *ante rem* perspective on structuralism proposes a dual ontology populated by places in structures and structures. Places are characterized only by structural relations; structures are characterized in a dim way, with a primitive identity criterion.

In a paper published in 2001, "The identity problem for realist structuralism", Keränen challenges *ante rem* structuralism by arguing that Shapiro's proposal fails in giving an adequate account of the identity of places in structures. The argument involves the idea that places in structures, as characterized by Shapiro, seem to violate Leibniz's Principle of Identity of Indiscernibles. Indeed, Keränen shows that, in the case of structures having non-trivial automorphisms (*i.e.*, not only an identity transformation), some places cannot be distinguished from each other. This is the case of the structure ′ of the integer numbers, where the places for numbers 1 and -1 should be identified according to Shapiro's criteria. If this were the case, the structure ′ would collapse into the structure ∞ of the natural numbers. Some authors consider that, if Keränen is right, then Shapiro's proposal clearly fails to the extent that it identifies entities that are mathematically different (MacBride 2005).

This problem triggered an interesting discussion between Keränen and Shapiro (Keränen 2006, Shapiro 2006a, 2006b), where both authors consider possible solutions to the identity problem for *ante rem* structuralism. Those solutions involve one of the following alternatives: either the rejection of Leibniz's Law and the assumption that places in structures are, in a certain sense, non-individuals, or the adoption of some kind of essentialism about places, which requires to admit some non-relational properties.

The debate has shown that the problem can hardly be solved in terms of Shapiro's theory of structures, since its formulation in second-order language with a set-semantics commits the theory with Leibniz's Law. Therefore, the only way to avoid the difficulties in this context is by accepting non-structural or non-relational properties; but this move would invalidate the whole structuralist proposal.

4. Two Domains, One Problem

In the previous sections, we have presented two independent problems: the problem of quantum indistinguishability and the problem of places in mathematical structures. In spite of the fact that both challenge Leibniz's Law, at first sight they seem to be independent and unrelated issues. Of course, this is the case when one considers the disciplines to which they belong: whereas one is rooted in an empirical science as physics, the other arises in mathematics, a formal science. Nevertheless, as we will see below, they are two instances of the same general problem. We will proceed by showing that they admit not only a common formulation, but also common possible treatments.

At the end of Section 2, we alluded to a third perspective to treat quantum indistinguishability, which belongs to the epistemological trend called "structural realism". Proposed for the first time by John Worrall (1989), structural realism aims at supplying an adequate answer to the pessimistic induction while retaining some of the intuitions considered right in the no-miracle argument: "[There] is [a] reasonable way to have the best of both worlds: to give the argument of scientific revolutions its full weight and yet adopt some sort of realist attitude towards presently accepted theories in physics and elsewhere –through structural realism, a position adopted by Poincaré, and here elaborated and defended." (Worrall 1989, p. 99). This position is formulated in the context of the debate between scientific realism and anti-realism, and it asserts that the parts of the past theories that are retained through theory change are related to the mathematical structure of a theory and not to its theoretical content.

Structural realism does not seem to have serious problems as an epistemic thesis: it asserts that we can only know the structural aspects of reality, which are what the mathematical structure of our scientific theories can reveal. According to Psillos (1999), this position is committed to the idea that scientific theories can reveal the structure of the unobservable world by means of their own mathematical structure; that mathematical equations –which are retained through scientific change– express real relations among entities (of which we can only know that they stand in these mathematically expressed relations to each other); and that different ontologies (different theoretical interpretations) may satisfy the same mathematical structure, but there is no reason to prefer one over any other.

Structural realism claims to be a realist proposal, and intends to differentiate itself from instrumentalism. Nevertheless, Psillos agrees with Laudan (1981) in the idea that the appeal to mathematical continuity, without further arguments, is not enough to defend a realist position. Here, we will not analyze the criticisms to structural realism. Let us just say that structural realism seems to be acceptable from an epistemic point of view, but it is insufficient when assessed from a metaphysical point of view, since it says nothing about existence. In fact, according to Worrall, knowledge is limited to structures, while knowledge about entities or procedures is impossible. Ladyman (1998) claims that a metaphysical version of structural realism is necessary. In this sense, he turns Worrall's epistemic position into an ontological thesis by offering such a version: if there is only knowledge about the structural aspects of reality, this is due to the fact that structures are all that there is. He proposes to replace the traditional ontology based on individuals, properties and relations with an ontology where structures are primitive and have ontological subsistence. Although inspired in structural realism, van Fraassen (2006) criticizes both formulations of this philosophical thesis, Worrall's and Ladyman's, and argues that the traditional perspectives have missed the point. By focusing his attention on quantum mechanics, van Fraassen claims that the theory only characterizes structures, which can be fulfilled with different kinds of items; however, which particular kind of items is selected for this task is completely irrelevant.

As we have seen, the ontological version of structural realism in the philosophy of science stresses that reality has some structural properties. Nevertheless, even accepting this viewpoint, one can still ask "What are those structures?" If structure is the fundamental category in the ontological domain, we should be able to provide a conceptual framework that allows us to talk about structures and the places defined by them. In other words, we need a consistent ontological characterization of structures and a consistent framework that allows us to understand the nature of structures and to conceive them as ontologically prior to their instances. When expressed in these terms, it is not hard to relate this requirement with the call for a consistent viewpoint about structures in Shapiro's *ante rem* structuralism. Indeed, the logical formalism proposed by Shapiro could be viewed as an answer to the claim of Ladyman (1998) on this subject.

However, the relation goes far beyond the agreement about the need of a consistent theory of structures. Let us focus now on the logical domain. As we have seen in Section 2, one of the two main answers to the problem of quantum indistinguishability is based on the denial of the individuality of quantum items. We have also seen that the development of a system of logic suitable for dealing with such non-individuals is an important goal to be achieved by the defenders of this position. The quasi-set theory (in his different formulations) proposed and developed by Newton da Costa, Steven French and Decio Krause (see Dalla Chiara, Giuntini & Krause 1998) is a concrete proposal in this sense: it is a meta-mathematical approach broader than Zermelo-Fraenkel's, deliberately formulated to treat quantum indistinguishability. This logical formalism distinguishes two kinds of entities, micro-objects and macro-objects, and explores the possibility of extending the basis of classical set-theory toward a theory where indistinguishability implies identity only for some entities, the macro-objects.

In the light of this consideration, let us recall the debates in the philosophy of mathematics: in Section 3, we have criticized Shapiro's proposal for its commitment with Leibniz's Law implicit in its relying on Zermelo-Fraenkel set theory and second order classical logic. Now it is clear that this problem can be solved by means of a formal tool developed in the context of a different philosophical discussion, that of quantum indistinguishability. Indeed, the quasi-set theory could be easily adapted from the physical realm to the mathematical domain by conceiving structures as macro-objects and places as micro-objects (see also the proposal of Krause 2005). On this basis, the idea of individual structures with a number of different but indistinguishable places would obtain the necessary formal framework.

Up to this point we have shown that both problems, that of quantum indistinguishability in the philosophy of physics and that of structural realism in the philosophy of mathematics, may be given the same logical solution. Until now, we have only considered the usual answers to those problems: the proposal of certain apparently *ad hoc* extra-mathematical or extra-physical properties, as well as the development of a theory of structures and of a logic for indistinguishable but different entities. In spite of their different nature, at last, all of these answers point to ontological matters. This is the problem we shall face in the next paragraphs: the possibility of conceiving a quantum ontology composed of indistinguishable but different subsystems, and a structural ontology composed of indistinguishable but different places.

5. A Single Ontological Solution

Following a traditional argument of Samuel Clarke in his debate with Leibniz, Max Black (1962) asserts that a completely symmetric universe, with two spheres "identical" in a physical sense (with the same non-dynamical properties), would be a universe with two different but indistinguishable spheres. Let us begin with noticing that the argument pre-supposes that the strongest version of Leibniz's Law is false. In fact, if the law were true, the conclusion of the argument could be rejected by rejecting one of its premises: in the universe there cannot be two objects that are intrinsically identical, that is, identical regarding their intrinsic properties.

But the Clarke-Black argument not only pre-supposes the falsity of the strongest version of Leibniz′s Law, but also challenges a weaker version of the principle, according to which it is not possible for two objects to be indiscernible regarding all of their properties and relations. In fact, in a completely symmetric universe that only contains those intrinsically identical spheres, say spheres a and b, all the relational properties would also be identical. Formally, $\forall R(Rab \leftrightarrow Rba)$. Consequently, according to that weaker version of Leibniz's Law, we have to conclude that a=b: both spheres are one and the same object with two different names. But since the universe with two intrinsically identical spheres is possible, physically speaking (see Earman 1986), and it is different from a universe with one single sphere, the Clarke-Black argument shows the contingency of the weaker version of Leibniz's Law. In turn, if one accepts the relational nature of space, also the weakest version of the principle, the one that includes space-time properties, falls under the strength of the Clarke-Black argument. If the spatial properties of the spheres are relational, they turn out to be symmetric and, hence, they do not distinguish between the spheres.

Let us now consider a new example: a universe with one single sphere. It seems evident that this example does not represent a challenge to Leibniz's Law, since we are dealing with a single object. But the situation changes when we conceptually "divide" the sphere into two perfectly identical halves and give a different name to each half. Now we are faced with the same case as that of the Clarke-Black argument: the two halves of the sphere are completely identical to each other regarding all of their intrinsic properties, and all their relational properties are symmetric. Therefore, they seem to violate Leibniz's Law as the two spheres did in the Clarke-Black example. But, does this example really represent a violation of Leibniz's Law? The answer depends on how the ontology is conceived. If the sphere is the legitimate object and its parts have no other ontological status than that of being an abstraction from the complete object, the answer is "no". If, on the contrary, the parts have ontological priority over the whole sphere, whose internal symmetry is the result of the relations of its parts, the answer is "yes".

In the case of the single sphere divided into two halves, one might consider that the parts are only abstractions with no independent existence. But this is not the case in any situation. Let us now consider a third example: a universe that only contains a rigid body, symmetric regarding a certain plane and whose behavior is ruled by classical mechanics. Again, we may decide between two alternatives. We can consider that there are two intrinsically indistinguishable objects separated by a constant distance, and that they violate Leibniz's Law; or we can consider that there is a single object, the rigid body, and then the principle does not apply. Does this example imply a violation of Leibniz's Principle? Again, the answer depends on how the ontology is conceived. If the rigid body (an iron bar, for example) is the legitimate object and its two halves have no other existence than that of being parts of the whole object, the answer is "no". If, on the contrary, the two halves of the body have ontological priority, and the symmetry of the rigid body results from the spatial disposition of those halves, the answer is "yes". But in this case, the same affirmative answer is obtained when we consider that the physical parts of the body (for instance, its molecules) have ontological priority: in this case, we are not dealing with abstract conceptual entities, but with very concrete space-time objects.

Any object with an intrinsic symmetry can be used to design these kinds of examples. The interpretation given in every particular case depends on the ontology accepted as primitive. Cases that do not seem problematic for Leibniz's Law from a certain viewpoint may turn out to violate even the weakest version of the principle when the ontology is conceived from a different perspective. The clue is to define which objects are legitimate objects and which objects are not.

When the ontological problems of *ante rem* realism are considered in the light of the above argument, it is easy to see that its difficulties regarding Leibniz's Law arise due to the fact that places themselves are considered as genuine objects. As we have seen, in Shapiro's version of structuralism, places are genuine objects, on the same footing as structures: this is its main obstacle. Shapiro's position forgets the ontological priority of structures and, hence, the ontological dependence of places from structures. If structures are the legitimately prior entities and they have internal symmetries (as it is indeed the case in structures with non-trivial automorphisms), the existence of different but indistinguishable places can be easily interpreted in analogy to the examples of the single sphere and the rigid body. Therefore, the difference between 1 and −1 in arithmetic would be a consequence of the symmetry of structure ′ of the integer numbers. The reason why 1 and −1 are not the same place is that the structure ′ that contains them is different from the structure ∞ that only contains natural numbers; the latter does not have non-trivial automorphisms, while the former does. It is on the basis of this difference in the properties of structures that the non-identity of places can be explained.

Although we have argued that the problems of quantum indistinguishability and of structural realism are actually two manifestations of the same problem, until now we have considered ontological issues in the mathematical realm. However, it is precisely that fact that allows us to apply the same strategy as that applied to the mathematical problem to the physical problem. If in the mathematical realm the ontological priority of structures over places supplies a way out of the difficulties of *ante rem* structuralism, in the quantum realm the clue is to admit the ontological priority of the composite system over the component subsystems. From this perspective, the indistinguishability of identical "particles" turns out to be, when analyzed in depth, no more than the statistical manifestation of the relationships among the subsystems of a symmetric composite system. This way of conceiving indistinguishability can be easily articulated with the modal-Hamiltonian interpretation of quantum mechanics (Lombardi & Castagnino 2008, Castagnino & Lombardi 2008, Ardenghi, Castagnino & Lombardi 2009, Lombardi, Castagnino & Ardenghi 2010, Lombardi, Fortin, Ardenghi & Castagnino 2011), which proposes an ontological view of quantum systems as bundles of properties. However, the complete development of this point is beyond the scope of the present chapter and will be developed in a future work.

6. Final Remarks

In this chapter we have considered and compared two problems that traditionally have not been related to each other. One of them belongs to the foundations of quantum mechanics, the other emerges in the philosophical foundations of mathematics. The first one results from empirical facts, the second is rooted in a particular philosophical position about mathematics. This means that, in a certain sense, the two problems come from different worlds. And their solutions may be conceived as completely independent: one could adopt one solution in one domain and a completely different solution in the other. Nevertheless, the two problems share very basic features since both involve structures and the relations among the elements of the structures. As a consequence, in both cases, the same alternative emerges: either the rejection of Leibniz's Law or the postulation of inaccessible properties. These agreements show that, from a general philosophical viewpoint, they are manifestations of a single ontological problem, which consists of understanding the nature of the items defined by structures and, even more basically, the nature of structures themselves. On this basis, in spite of the fact that they belong to different domains, if a new solution for one of the problems were found, one can expect that it would also work for the other problem: it would be a splendid opportunity to kill two birds with one stone.

Acknowledgments

We want to thank all of the members of the Buenos Aires Group on Philosophy and Foundations of Physics and Chemistry, to which we belong, for supplying us the necessary intellectual context to develop these ideas. This work has been supported by grants of CONICET, ANPCyT, and UBA, Argentina.

References

Ardenghi, J. S., Castagnino, M. & Lombardi, O. (2009). "Quantum mechanics: modal interpretation and Galilean transformations." *Foundations of Physics*, 39: 1023-1045.

Ballentine, L. (1998). *Quantum Mechanics. A Modern Developement*. Singapur: World Scientific.

Benacerraf, P. (1965). "What numbers could not be?" *The Philosophical Review*, 74: 47-73.

Black, M. (1962). "The identity of indiscernibles." *Mind*, 61: 153-64.

Castagnino, M. & Lombardi, O. (2008). "The role of the Hamiltonian in the interpretation of quantum mechanics." *Journal of Physics. Conferences Series*, 28: 012014.

Castellani, E. (ed.) (1998). *Interpreting Bodies. Classical and Quantum Objects in Modern Physics*. Princeton: Princeton University Press.

da Costa, N. C. A. & Krause, D. (1994). "Schrödinger logics." *Studia Logica*, 53: 533-550.

da Costa, N. C. A. & Krause, D. (1997). "An intensional Schrödinger logic." *Notre Dame Journal of Formal Logic*, 38: 179-194.

da Costa, N. C. A. & Krause, D. (1999). "Set-theoretical models for quantum systems." In M. L. Dalla Chiara, R. Giuntini & F. Laudisa (eds.), *Language, Quantum, Music*. Dordrecht: Kluwer, 171-181.

Dalla Chiara, M. L., Giuntini, R. & Krause, D. (1998). "Quasiset theories for micro-objects: a comparison." In E. Castellani (ed.), *Interpreting Bodies. Classical and Quantum Objects in Modern Physics*. Princeton: Princeton University Press.

Earman, J. (1986). *A Primer on Determinism*. Dordrecht: Reidel.

French, S. & Krause, D. (2006). *Identity in Physics: A Historical, Philosophical, and Formal Analysis*. Oxford: Oxford University Press.

French, S. & Redhead, M. (1988). "Quantum physics and the identity of indiscernibles." *The British Journal for the Philosophy of Science*, 39: 233-246.

Hellman, G. (1996). "Structuralism without structures." *Philosophia Mathematica*, 3: 100-123.

Hesse, M. (1963). *Models and Analogies in Science.* London: Sheed and Ward.

Huggett, N. (1995). "What are quanta, and why does it matter?" *PSA 1994* (*Proceedings of the 1994 Biennial Meeting of the Philosophy of Science Association*) Vol. 2. East Lansing: Philosophy of Science Association, 69-76.

Jauch, J. (1968). *Foundations of Quantum Mechanics.* Reading Mass.: Addison Wesley.

Keränen, J. (2001). "The identity problem for realist structuralism." *Philosophia Mathematica*, 9: 308-330.

Keränen, J. (2006). "The identity problem for realist structuralism II: A reply to Shapiro." In F. MacBride (ed.), *Identity and Modality*. New York: Oxford University Press, 146-163.

Krause, D. (2005). "Structures and structural realism." *Logic Journal of IGPL*, 13: 113-126.

Krause, D. (2006). "Einstein y la indiscernibilidad." *Praxis Filosófica, Nueva Serie*, 22: 113-130.

Krause, D. (2010). "Logical aspects of quantum (non-) individuality." *Foundations of Science*, 15: 79-94.

Ladyman, J. (1998). "What is structural realism?" *Studies in History and Philosophy of Science*, 49: 409-424.

Ladyman, J. & Bigaj, T. (2010). "The principle of the identity of indiscernibles and quantum mechanics." *Philosophy of Science*, 77: 117-136.

Laudan, L. (1981). "A confutation of convergent realism." *Philosophy of Science*, 48: 19-49.

Lombardi, O. & Castagnino, M. (2008). "A modal-Hamiltonian interpretation of quantum mechanics." *Studies in History and Philosophy of Modern Physics*, 39: 380-443.

Lombardi, O., Castagnino, M. & Ardenghi, J. S. (2010). "The modal-Hamiltonian interpretation and the Galilean covariance of quantum mechanics." *Studies in History and Philosophy of Modern Physics*, 41: 93-103.

Lombardi, O., Fortin, S., Ardenghi, J. S. & Castagnino, M. (2011). *Introduction to the Modal-Hamiltonian Interpretation of Quantum Mechanics*. New York: Nova Science Publishers.

MacBride, F. (2005). "Structuralism reconsidered." In S. Shapiro (ed.), *The Oxford Handbook of the Philosophy of Mathematics and Logic*. New York: Oxford University Press, 563-89.

MacBride, F. (ed.) (2005). *Identity and Modality*. New York: Oxford University Press.

Post, H. (1963). "Individuality and physics." *Listener*, 70: 534-537.

Psillos, S. (1999), *Scientific Realism. How Science Tracks Truth*. London: Routledge.

Redhead, M. & Teller, P. (1992). "Particle labels and the theory of indistinguishable particles in quantum mechanics", *The British Journal for the Philosophy of Science*, 43: 201-218.

Resnik, M. (1997). *Mathematics as a Science of Patterns*. Oxford: Oxford University Press.

Shapiro, S. (1997). *Philosophy of Mathematics. Structure and Ontology*. New York: Oxford University Press.

Shapiro, S. (2006a). "Structure and identity." In F. MacBride (ed.), *Identity and Modality*. New York: Oxford University Press, 109-145.

Shapiro, S. (2006b). "The governance of identity." In F. MacBride (ed.), *Identity and Modality*. New York: Oxford University Press, 164-173.

Shapiro, S. (ed.) (2005). *The Oxford Handbook of the Philosophy of Mathematics and Logic*. New York: Oxford University Press.

Singh, V. (2005). "Einstein and the quantum." *Current Science*, 89: 2101-2112.

Teller, P. (1983). "Quantum physics, the identity of indiscernibles and some unanswered questions." *Philosophy of Science*, 50: 309-319.

Teller, P. (1995). *An Interpretive Introduction to Quantum Field Theory*. Princeton: Princeton University Press.

Teller, P. (1998). "Quantum mechanics and *haecceities*." In E. Castellani (ed.), *Interpreting Bodies. Classical and Quantum Objects in Modern Physics*. Princeton: Princeton University Pres, 114-141.

van Fraassen, B. (1998). "The problem of indistinguishable particles." In E. Castellani (ed.), *Interpreting Bodies. Classical and Quantum Objects in Modern Physics*. Princeton: Princeton University Press, 73-92.

van Fraassen, B. (2006). "Structure: Its shadow and substance." *The British Journal for the Philosophy of Science*, 57: 275-307.

Worrall, J. (1989). "Structural realism: The best of both worlds?" *Dialectica*, 43: 99-124.

In: Philosophy of Science
Editors: S. Pintuck and C. Reynolds
ISBN: 978-1-62100-276-5

Chapter 4

THE "HOW DOES IT KNOW?" PROBLEM: A CONCEPTUAL PARALLEL BETWEEN HEISENBERG'S AND BOHM'S INTERPRETATIONS OF QUANTUM MECHANICS

James Henderson
University of Pittsburg at Titusville, Pennsylvania, US

1. INTRODUCTION

It is undeniable that quantum mechanics has enjoyed nothing less than staggering success in almost every way that scientific theories are judged. Although it is a theory whose laws govern the unimaginably small, it has had a profound impact on the everyday lives of people. Paul Davies says

> ...the quantum theory is primarily a practical branch of physics.... It has given us the laser, the electron microscope, the transistor, the superconductor and nuclear power. At a stroke it explained chemical bonding, the structure of the atom and nucleus, the conduction of electricity, the mechanical and thermal properties of solids, the stiffness of collapsed stars and a host of other important physical phenomena... These days, it is applied in many routine practical ways in engineering. (1983, p. 101)

The praise for the theory does not stop there. David Albert states that "[the principles of quantum mechanics] are the most precise mechanism for predicting the outcomes of experiments on physical systems ever devised. No exceptions to them have ever been discovered. Nobody expects any" (1993, p. 36). Going even further, if such is possible, Steven Weinberg writes that "[t]he one part of today's physics that seems… likely to survive unchanged in a final [physical] theory is quantum mechanics" (1993, p. 211). In this atmosphere of such high praise, it is ironic in the extreme to find that the canonical interpretation of the theory is routinely held to be a failed attempt to describe reality. Even though the Copenhagen interpretation is very likely still the guiding dogma of the average working physicist, serious students of the foundations of quantum mechanics rarely defend the standard formulation (Albert, 1994, p. 67).

In trying to make sense of queer lab results popping up all around the world, David Bohm published his Pilot-Wave interpretation in 1952. Bohm constructs a model that denies many of the counter-intuitive claims of the received view and avoids some of the weirdness of the Copenhagen interpretation, but a certain amount of it is forced on him by the phenomena to be explained– the theory itself is weird, as evidenced by Neils Bohr's now-famous line, "Anyone who is not shocked by quantum mechanics has not understood it" (quoted in Davies, 1983, p. 100); any interpretation of it must be strange to one degree or another. After a discussion of these phenomena, I will examine these two interpretations. In particular, I will consider a problem common to the two of them that I call the "How does it know?" problem. While it has been well known for some time that this plagues Heisenberg's Copenhagen interpretation, I will argue that it is likewise a concern for Bohm, though it appears in a different guise.

2. SubatomicPhenomena and Quantum Mechanics

The experimental results for which quantum mechanics must account are mind bending, but the most mysterious concept to come out of the theory is that of "superposition." To illustrate this idea in as straightforward and non-technical a fashion as possible, I will consider a sequence of simplified Stern-Gerlach experiments, culminating in an idealized version of the famous two-slit experiment. At the end of this discussion will be a clear statement of what

I call the "How does it know?" problem (the presentation here is based on that of [Albert, 1993, pp. 1-16]), and I will argue that Bohm's interpretation of quantum theory not so much solves this problem as relocates it. That is, it will be seen that Heisenberg and Bohm have analogous problems in their respective accounts.

It is a matter of empirical fact that of all electrons ever encountered in the lab or in the wild, statistically half of them have a spin value, when measured along the x-axis, of measure "left," and the other half have an x-spin value of "right" (I will discuss x-spin measurements in terms of left-right and y-spin measurements in terms of up-down for purposes of clarity). Simple, accurate machines (Stern-Gerlach devices) can be made to carry out this measurement. These measurements are repeatable; that is, when an electron has been measured to have an x-spin value of, say, "left," a subsequent test of the electron will yield the same result (given that the electron has not been disturbed in the meantime). Similarly, spin along the y-axis may be accurately measured. Statistically half of the electrons ever encountered in labs or elsewhere have y-spin "up," and half have y-spin "down." These measurements are also repeatable and easy to carry out.

Suppose one is interested in finding out if there is any correlation between the measures of x-spin and y-spin of electrons; i.e., given that the x-spin of an electron has a particular value, does this tell us anything about its y-spin value? It is easy to demonstrate (and has in fact been demonstrated) that precisely half of all electrons with x-spin "left" have y-spin "up" and half have y-spin "down." Similarly, half of all x-spin "right" electrons have y-spin "up" and half have y-spin "down." That is to say, there is no correlation between the x- and y-spin values.

Interestingly, if one measures the x-spin value of an electron (say it comes out "right") and then measures its y-spin value (say it comes out "up"), a subsequent measure of the x-spin value has a 50% chance of coming out "left" and a 50% chance of resulting in a value of "right." No way of determining which electrons will have their spin values changed and which will have theirs remain constant has ever been devised. Further, though many different kinds of measuring apparatuses built according to a variety of physical principles have been constructed, the randomization of x-spin by a measurement of y-spin has, in every test, always remained complete. That is, x-spin "left" or "right" is always precisely a 50-50 proposition – this is important; it implies that the randomization is not an artifact of the designs of the particular devices used to carry out the measurements. On top of this, Renninger-type experiments, measurements where the measured electron is *not* passed through a Stern-Gerlach device but simply bounced off mirrors, have been carried out, all with precisely the same statistical results. Spin randomization seems to have nothing whatsoever to do with the particular design of the measuring apparatus. It should be noted here that, just as a measure of y-spin randomizes the x-spin value, a measurement of x-spin also randomizes y-spin value.

Now consider a group of x-spin "left" electrons being fed into a y-spin measuring device. No matter which way the electrons leave the device, they are bounced, by way of reflecting walls (mirrors, essentially) that change nothing more than the direction of the electron's flight (this can be demonstrated) into a feeder that directs them into another x-spin measuring apparatus (the feeder, like the mirrors, changes nothing more than the trajectory of the electrons, and this can also be demonstrated). As the electrons in question have just passed through a y-spin measuring device, it is reasonable to expect the electrons to come out half "left" spinning and half "right" spinning with respect to their x-axes, but this is not what happens. All of the electrons come out with an x-spin value of "left" (sharp-eyed readers will have noted this experimental set-up is equivalent to the famous "two-slit" experiment).

Next, consider the same experimental set-up with an electron-stopping wall inserted into, say, the y-spin "down" path to the feeder. It seems reasonable to guess that only half of the electrons will reach the final x-spin device (as half will be blocked by the wall) and have their x-spin values measured. This is precisely what happens. The distribution of the final spin values is strange, however: precisely half (statistically speaking) have x-spin "left" and half have x-spin "right."

I will now directly address the idea of superposition. One may ask which path to the feeder the electrons took in the original two-path experiment (the one without the wall). They cannot have taken the y-spin "up" path because we know by experiment that the electrons taking that path have randomized x-spin values (recall that in the original two-path experiment the electrons were uniformly of x-spin "left"). They cannot have taken the y-spin "down" path since the wall could just have easily been placed in the y-spin "up" path, and our previous argument would hold. They cannot have taken neither path, since if we block both paths no electrons get through at all (this, too, can be demonstrated). They cannot have taken both paths, since when we check on them in mid-flight each electron is located on one path or the other. In summation, they have not taken one path, they have not taken the other, they have not taken neither, and they have not taken both. In spite of all this, they do in fact get to the final x-spin device. Physicists say that in cases like this the electrons are in a superposition of taking the two paths – not one, not the other, not both, and not neither. This is a fancy way of saying they get there but we don't know how.

It gets stranger still. Since an electron having y-spin "up" amounts to its taking the y-spin "up" path when it is sent through a y-spin device and an electron's having y-spin "down" amounts to its taking the y-spin "down" path when it is sent through a y-spin apparatus, the electrons being in a superposition of taking upper and lower paths means that they are in a superposition of being in y-spin "up" and y-spin "down" states: that is, they don't have y-spin "up," y-spin "down," both y-spin "up" and y-spin "down," nor do they have neither y-spin "up" nor y-spin "down."

As strange as this seems, there is an algorithm that accounts for all the bizarre behavior (like superposition) of quantum entities. This algorithm, of course, is quantum mechanics. Physical states of systems may be described with a mathematical entity known as a wave function. Wave functions (and their corresponding states of affairs) evolve in a continuous, deterministic fashion according to Schrodinger's equation. Using quantum mechanics, we may make predictions as to how experiments will come out; the quantum mechanical algorithm gives us the percentages associated with any particular outcome. Since experiments are repeatable, there must be a mechanism that changes, say, a 50-percent likelihood of an up-spinning outcome and a 50-percent likelihood of an down-spinning outcome into a 100-percent certainty of an up-spinning result after an initial measurement of up-spin. This mechanism is the collapse postulate. The collapse postulate, discontinuous and probabilistic, says, bluntly, that when experiments are made, definite results occur. After a measurement is made, the wave function collapses from a state which says "there is a 50-percent chance of an up-spinning result and a 50-percent chance of a down-spinning result if an experiment is made" to a state saying "the electron is up-spinning." Naturally, had the experiment yielded a result of down-spinning, the wave function would collapse to a description of a down-spinning electron.

3. THE COPENHAGENINTERPRETATION

Don Howard (2004) traces the evolution of the Copenhagen interpretation. What is now known as the Copenhagen Interpretation is largely the invention of Werner Heisenberg, not, as commonly thought, Niels Bohr. The term "Copenhagen Interpretation" itself was not coined until 1955, and then by

Heisenberg (Howard, 2004, p. 675). It differs substantially from Bohr's view, which took complementarity and entanglement to be the characteristic elements of quantum theory. It is Heisenberg who places collapse at the center of the Copenhagen interpretation, and in fact "Bohr never mentioned wave-packet collapse" (Howard, 2004, pp. 670, 1) and "always criticized Heisenberg for promoting the disturbance [collapse] analysis" (Howard, 2004, p. 672).

Though Heisenberg introduced "Copenhagen Interpretation" into the lexicon in the mid-1950s, conflation of his views with Bohr's began much earlier. As early as the 1930s "physicist friends of logical empiricism, such as Philipp Frank… and Pascual Jordan… cited Heisenberg on the role of the observer in order to claim quantum mechanics as yet another scientific vindication of verificationism" (Howard, 2004, p. 677). Jordan wrote that positivism was the viewpoint of Heisenberg and Bohr, but quoted only Heisenberg in support of this claim (Howard, 2004, p. 677). Therefore the dominant view was extant before it was given its current moniker.

It was for different reasons that others, Bohmincluded, continued presenting Heisenberg's views as representative of the orthodox view (Howard, 2004, p. 677). "Bohm seized upon Heisenberg's version of a Copenhagen interpretation for legitimating his hidden variables program" by contrasting it with what he took as weaknesses in Heisenberg's account (Howard, 2004, p. 678). Given this, it is ironic that Bohm's theory is troubled by a problem directly analogous to one bedeviling Heisenberg's.

In any case, given the quantum mechanical algorithm, there are many ways to interpret it. The Copenhagen interpretation is the standard way of thinking of the quantum theory and is given in all physics textbooks. It is hardly surprising that an interpretation of a theory that predicts such strange results is, at best, counter-intuitive.

The basic assumptions of the standard interpretation of quantum mechanics may be stated succinctly:

1) The wave function with its probability interpretation determines the most complete possible specification of the state of an individual system.
2) The process of transfer of a single quantum from observed system to measuring apparatus is inherently unpredictable, uncontrollable, and unanalyzable. (Bohm, 1983, p. 372)

It doesn't take long, however, for intuition to take a real beating. In these two assumptions lies the kernal of the Heisenberg Uncertainty Principle. This places a definite limit on the precision with which we may consider x-axis spin and y-axis spin (or any pair of incompatible quantities such as position and momentum or time and energy) as simultaneously existing properties of a given particle. That is, the standard interpretation of the quantum theory takes very seriously the idea of superposition: If an electron has a particular momentum, then there is no fact of the matter concerning the particle's position; it is not here, not there, not both here and there, and not neither here nor there. The Copenhagen interpretation holds that asking for the position of this particular electron is a category mistake because the electron simply isn't located anywhere.

Stranger things follow. If superpositions of states are taken to be representations of circumstances in the real world, we must consider what happens when a measurement occurs. If we measure the x-axis spin of an electron with well-defined y-axis spin, there is no matter of fact about the x-axis spin of the particle until the measurement is made. However, the instant a measurement is made the x-axis spin is either left or right, each possibility occurring with 50-percent likelihood. The wave function collapses in a completely probabilistic manner even though we know that when measurements are not being made the wave function evolves *deterministically* according to Schrodinger's equation. What is it about measurements that alter the way subatomic particles behave? Why should matter behave in one way when experiments are being done but in a fundamentally different way when they are not being carried out? Aren't experimenters and their equipment made of the same kind of particles being observed? If they are (and they are), there is a (complex) wave function that describes experimenter, equipment, and observed system. This wave function must evolve deterministically until it is measured, and until that point the experimenter, her equipment, and the observed system are in a superposition of states of having the result of the experiment coming out in one way or another (x-spin left or x-spin right). But when we measure *that* system (and collapse it), another more complex wave function describing observer, equipment, system, and meta-observer now describes a superposition of events. While this process of observing observers may be backed up indefinitely, we take note of the fact that this is not what we experience in the lab – experiments do have definite results. Again, what is it about measurements that alter the way subatomic particles behave? Does it necessarily involve humans? Must records be created? Are worlds, identical to our own until the time of the observation, being created? For the purposes of this essay, I will concentrate on the following question: How do the measured entities "know" we are watching? That is, how do electrons (or protons, or whatever entities are involved) know a measurement is taking place so that they may change their behavior accordingly?

It is at this stage that adherents to the standard interpretation must make the claim that the act of measurement is a very special process indeed. Now measurement is a very slippery concept, and no one has been able to arrive at an explication of it that satisfies everyone. But whatever else it is, a measurement seems to be a process that causes wave functions to collapse. If this doesn't seem like much of a solution, it at least makes clear why so many philosophers are troubled by the way the Copenhagen school deals with the problem involving observers and physical systems.

4. BOHM'SNEW PILOT-WAVEINTERPRETATION

In 1952, David Bohm published his own interpretation of quantum mechanics, an updated version of de Broglie's Pilot-wave interpretation. Though virtually ignored by practicing physicists, it is a realistic account of the strange results of experiments on subatomic particles while maintaining all of the theory's accuracy with regard to predictions of experimental outcomes. Bohm's motivations are made clear in his article:

> The usual interpretation of the quantum theory... requires us to give up the possibility of even conceiving precisely what might determine the behavior of an individual system at the quantum level, without providing adequate proof that such a renunciation is necessary. (1983, p. 371)

Bohm's objection that Heisenberg's interpretation gives no explanation of why particles behave as they do calls into question Davies' statement that quantum mechanics "at a stroke" explained chemical bonding, the structure of the atom and nucleus, or, for that matter, anything. Bohm's theory is free of superpositions, special status for observers, and the on-again off-again existence of position, momentum, and other qualities of subatomic particles (particles always have a well-defined position for Bohm). He goes so far as to say that there may be adjustments to his theory to explain unanticipated particle behavior that is unavailable to the standard interpretation.

Bohm begins by postulating that the quantum mechanical wave function of a particle describes an actual force field (a "psi-field") that evolves in the manner prescribed by the Schrodinger equation. "This field exerts a force on the particle in a way that is analogous to... the way in which a magnetic field exerts a force on a charge" (Bohm, 1983, p. 373). He also postulates the existence of a quantum mechanical potential, a quantum mechanical analog of the classical potential. The quantum mechanical potential, along with the classical potential and the psi-field, determines the motion of the electron. If the momentum **p** of a particle is restricted by the equation

$$\mathbf{p} = \nabla S(\mathbf{x}),$$

and if we assume that the (well-defined) position of the particle in question is given statistically by the intensity of the psi-field, all the familiar results of quantum mechanics will be preserved. To be clear: The Copenhagen interpretation states that the intensity of the wave function gives the probability that the particle will be found in a particular region if a measure of position is carried out; Bohm believes the probability of the particle's position is given by the intensity of the psi-field at the point of interest. While for Heisenberg the particle does not have a position until the measurement is carried out, Bohm contends that the position is always defined, but it is an epistemic shortcoming that requires us to give its position only probabilistically through the wave-function. Only upon the completion of a measurement of its position is the exact location of the particle known.

We now consider how Bohm's theory explains the Stern-Gerlach experiments examined earlier. Assume the initial position of an up-spinning particle is in the upper half of the region where the intensity of the psi-field is non-0 (a 50-50 proposition). The particle, along with the psi-field, enters the x-spin measuring Stern-Gerlach device. The right-spin portion of the wave function moves out through the right-spin aperture of the device (for clarity's sake, assume it is the upper aperture) while the left-spin portion exits through the left-spin aperture (the lower opening). As long as the particle is in the region where the different parts of the psi-field overlap it will continue to travel in a straight line. However, when the particle is in the region occupied exclusively by the right-spin portion of the field it will travel along an upward path through the right-spinning aperture. But now the particle could be located anywhere in the region where the intensity of the right-spin portion of the psi-field is non-0. Therefore, it is 50-percent likely to exit through the up-spin aperture and 50-percent likely to exit through the down-spin aperture when it arrives at the y-spin measuring device. Its position in the right-spin portion of the psi-field will determine through which aperture in the y-spin measuring device it leaves and, thus, the outcome of the y-spin measurement.

This is why the measure of horizontal spin has randomized the value of its vertical spin.

When we reunite the halves of the wave function as in the second experimental set-up (the non-blocked experiment involving the feeder), we again see the familiar quantum mechanical result. Assume the right-spinning particle has initial position in the lower half of the psi-field. It exits through the down-spinning aperture (assuming that is the lower one) and is deflected into the feeder where it is reunited with the up-spinning portion of its wave function (which has taken the upper route to the feeder). Now the particle is exactly as it was before it entered the y-spin measuring device, so it is necessarily right-spinning again.

Now we reconsider the puzzling "blocked path" experiment in Bohmian terms. Since there is a 50-percent chance the electron will be located in the lower half of the psi-field, half of the electrons will be stopped at the wall. The other half will proceed along the upper path, but the down-spinning portions of the psi-fields don't make it to the feeder. Now we have an up-spinning electron which is equally likely to be located in the top or bottom half of its psi-field; hence, it is equally likely to exit the right- or left-spin exits of the final x-spin measuring device.

There is an unexpected result of taking the Bohmian position. Imagine a left-spinning electron located in the top half of its psi-field approaching a y-spin measuring device. It then follows that the electron should exit through the up-spin exit (if that is the upper one). But if the box is inverted prior to the electron's entry (so that now the down-spin exit is on top), the electron will come out down-spinning. On this account electron spin is not an attribute of electrons – an electron's only attribute is its position.

We now consider EPR-type experiments involving instantaneous action at a distance. Consider a non-separable, two-particle system that is stationary. If the particles are accelerated in opposite directions, we know that their y-axis spins (when measured) will be opposite – one up, one down. The standard interpretation holds that the spin measurements are not well defined until a measurement is made, but at that point they will instantaneously take on opposite values no matter how far apart the electrons are. No explanation other than the collapse of the wave function is possible. Bohm argues that as the measured electron enters the y-spin measuring device, the psi-field begins to split. Instantly the psi-field of the unmeasured electron also begins to part. The moment the measured electron is located in a region where only the intensity of the (say) up-spin portion of the psi-field is non-0, the unmeasured electron moves into a region where only the down-spin psi-field has positive intensity. This is very important for Bohm. Since he claims that the psi-field is a real force field, it must behave in a continuous manner; there can be no discontinuous "collapse." Bohm's theory is non-local (as it must be if it is to preserve the results of the quantum theory), but it is non-local in a concrete way. Force fields and particles move in concert instantaneously at a distance, but there is no "jumping" of states of the system.

Has Bohm accomplished his goal of providing a superior interpretation of quantum mechanics? He has certainly avoided many of the problems associated with the Copenhagen school of thought. A discontinuous collapse of the wave function is not a concern for Bohm. He has orchestrated what Albert calls an "effective collapse" (1993, p. 162) by splitting his psi-field in a smooth, continuous manner when measurements are made, thus avoiding concerns regarding a conscious observer precipitating a change in the behavior of matter. He has succeeded in naming a cause for the randomization of one observable quantity by measurement of another (e.g., randomizing y-axis spin by measuring spin along the x-axis). In his scheme this is a function of the position of the particle within its psi-field (which we may only know probabilistically). Further, his ontology has no need for special processes (measurement) or special beings (observers).

Should we rush to adopt the long overlooked theory? It is not without its own problems. While Bohm has no need of special status for measurements or observers, his theory does postulate some non-standard entities. The wave function is no longer a description of a particle or system; rather the psi-field is regarded as a force which, with the classical and quantum mechanical potential (itself a non-standard element of Bohm's ontology) pushes particles around. Bohm needs real forces to supply causes for the actions of subatomic particles, but the psi-field he invokes is a strange force indeed.

> Bohm's theory entails that the 'empty' part of the wave function... is completely undetectable. One of the consequences of [the theory] is that only the part of any given particle's wave function that is currently occupied by the particle itself can have any effect on the motions of other particles. (Albert, 1994, p. 66)

I will now set up one last Stern-Gerlach experiment to see how this may expose the "How does it know?" problem in Bohm's model.

A right-spinning electron (e_1) is fed into a y-spin measuring device. It exits through the up-spin aperture with the up-spin portion of its psi-field (we may have to do the experiment several times to get this result) and hits an electron detector (a phosphour plate) which lights up when struck (we will know if the electron takes the lower path since the detector will stay dark) while the down-spin portion of the psi-field takes the lower path to a feeder. When the detector lights up, another up-spinning electron (e_2) is fired into the feeder from an emitter so as to arrive simultaneously with the down-spin portion of the psi-field of e_1. The feeder sends both into a waiting x-spin measuring device. e_2 will exit the right-spin aperture with 50 percent likelihood or the left-spin aperture with 50 percent likelihood (both theories predict this). Compare this with the non-blocked experimental set-up involving the feeder, in which the electron exited the final x-spin measuring device necessarily through the right-spin aperture. Heisenberg has an easy explanation for the difference: Nothing but an up-spinning electron (e_2) entered the final x-spin measuring device, so it will necessarily have a 50-50 chance of coming out with right-spin (or left-spin); e_1's wave function doesn't come into play because it is a description of e_1, not e_2. Bohm must take a different approach. He must claim e_1's psi-field has no effect on e_2 because it is the psi-field of a different electron (e_1). That is, Bohm must say the psi-field taking the lower path is the *forceof* e_1, not e_2. How can we have a force *of an electron*? All electrons are the same.

> Every electron we have ever encountered... is found to be identical. All have the same electric charge, the same spin, the same mass, to the accuracy of measurement. They all behave in the same way in interactions with other particles. (Barrow, 1991, p. 266)

How is it, then, that the e_1 and e_2 behave differently when e_1's psi-field is exposed to each? When conducting EPR-type experiments, it is not uncommon to again hear the famous question "How does the unmeasured particle 'know' the outcome of the measurement performed on the other electron?" Here we may ask, "How does the psi-field 'know' which particle it is exposed to when the particles are identical in every way?"

One way out of this morass may be to say that when the position of the electron is determined (this occurs when e_1 strikes the electron detector), any non-occupied portion of the electron's psi-field (that is, not occupied by its associated electron) disappears – it simply winks out of existence. The problem with this "solution" is that there is nothing in the formalism that tells us this should be the case. The addition of such a plank into Bohm's platform would be a completely *ad hoc* move forced on him by observation, but, in fairness, the same could be said of the addition of the collapse postulate to the Copenhagen interpretation.

When we consider EPR-type experiments, Bohm fails to give us the one thing he strives for most: causation. When the psi-field of the tested electron begins to split, how is it that instantaneously and at a distance the second psi-field begins its split? How does the unmeasured psi-field "know" to push its electron to its appropriate destination? When the measured electron moves into a region where, say, only the up-spin portion of the psi-field has non-0 intensity, the unmeasured partner must move to a region where only the intensity of the down-spin portion of the psi-field is non-0. How does the psi-field instantly "know" that this is the appropriate destination?

We now turn to another reason Bohm thinks his theory should be adopted. Bohm has restricted the location of particles to a statistical distribution given by the intensity of the psi-field. He has also, as has been stated earlier, restricted particle momentum to

$$\mathbf{p} = \nabla S(\mathbf{x}).$$

In Bohm's theory, however, these restrictions are "not inherent in the conceptual structure" (Bohm, 1983, p. 373). In 1952, technology did not allow measurements involving distances less than 10^{-13} cm. If errors were found in the quantum theory's predictions involving distances less than this quantity, Bohm's theory could be adjusted to fit the data without damaging the theory itself. That is, Bohm could ease these restrictions slightly to fit newfound data in such a way as to produce only "negligible effects in the atomic domain" (Bohm, 1983, p. 374). The standard interpretation, with its insistence that the wave function is the most complete description of a system possible, does not have this flexibility. Ignoring that flexibility is something Steven Weinberg finds objectionable in theories (Weinberg, 1993, p. 135), we may note that even in the face of advancing technology, no such shortcomings have arisen regarding the predictions of quantum mechanics. There is no guarantee that errors will never be found, but it is remarkable that in an age when we probe tiny distances looking for extraordinarily small fluctuations caused by passing gravitons, none of the straying from data Bohm so confidently expected has been discovered.

5. CONCLUSION

Bohm has given us a realistic alternative to quantum mechanics, but not in such a completely satisfactory way as he hoped. He avoided some of the perils of the standard interpretation, but only at a cost of stumbling into new ones. A real test of his theory is whether or not it may be extended to a form that is compatible with Einstein's general theory of relativity. Sadly, no really significant progress has been made on this front, but no one has succeeded in generalizing the standard interpretation in such a manner either.

Quantum mechanics has accounted for a multitude of experimental results in the sense that it supplies accurate predictions, but it has given no explanations of them. The task of creating a realistic successor to the quantum theory involves not only creating a relativizable theory, but also outlining causes for what is seen in the lab. While Bohm has made some headway, the scientific and philosophical worlds must continue to wait for the job to be finished.

REFERENCES

Albert, D. (1993). *Quantum Mechanics and Experience.* Cambridge: Harvard University Press.

_____. (1994). "Bohm's Alternative to Quantum Mechanics," Scientific American, May 1994, pp. 58-67.

Barrow, J. (1991). *Theories of Everything.* New York: Random House.

Bohm, D. (1983). "A Suggested Interpretation of the Quantum Theory in Terms of 'Hidden' Variables I," in *Quantum Theory and Measurement,* eds. J. Wheeler and W. Zurek. Princeton: Princeton University Press, 1983, pp. 369-82.

Davies, P. (1983). *God and the New Physics.* New York: Simon and Schuster.

Howard, D. (2004). Who invented the "Copenhagen interpretation"? A study in mythology, *Philosophy of Science* 71 (2004), pp. 669-682.

Weinberg, S. (1993). *Dreams of a Final Theory.* New York: Random House.

INDEX

A

B

C

D

E

F

G

H

I

J

K

L

M

N

O

P

Q

R

S

T

U

V

W

Y

Z